AGEING AND PLACE

In recent decades, population ageing has combined with expanding forms of health care and accommodation so that older people's lives are negotiated in, transformed and represented by an ever-increasing range of settings. In this context, *Ageing and Place* examines the many ways in which place influences older people's lives. The collection of reviews range from spatial macro-scale perspectives to considering relationships with and within specific settings. The anthology is written from a predominantly geographical viewpoint but also draws upon psychological and sociological approaches. *Ageing and Place* therefore spans research traditions and international perspectives, presenting a comprehensive review of research.

This innovative work is essential reading for a variety of postgraduate and undergraduate students, health professionals and social science researchers.

Gavin J. Andrews is an Associate Professor at the Faculty of Nursing, University of Toronto.

David R. Phillips is a Professor of Social Policy at Lingnan University, Hong Kong.

ROUTLEDGE STUDIES IN HUMAN GEOGRAPHY

This series provides a forum for innovative, vibrant and critical debate within Human Geography. Titles will reflect the wealth of research which is taking place in this diverse and ever-expanding field. Contributions will be drawn from the main sub-disciplines and from innovative areas of work which have no particular sub-disciplinary allegiances.

1 A GEOGRAPHY OF ISLANDS
Small island insularity
Stephen A. Royle

2 CITIZENSHIPS, CONTINGENCY AND THE COUNTRYSIDE
Rights, culture, land and the environment
Gavin Parker

3 THE DIFFERENTIATED COUNTRYSIDE
Jonathan Murdoch, Philip Lowe, Neil Ward and Terry Marsden

4 THE HUMAN GEOGRAPHY OF EAST CENTRAL EUROPE
David Turnock

5 IMAGINED REGIONAL COMMUNITIES
Integration and sovereignty in the global south
James D. Sidaway

6 MAPPING MODERNITIES
Geographies of central and eastern Europe 1920–2000
Alan Dingsdale

7 RURAL POVERTY
Marginalisation and exclusion in Britain and the United States
Paul Milbourne

8 POVERTY AND THE THIRD WAY
Colin C. Williams and Jan Windebank

9 AGEING AND PLACE
Perspectives, policy, practice
Edited by Gavin J. Andrews and David R. Phillips

10 GEOGRAPHIES OF COMMODITY CHAINS
Edited by Alex Hughes and Suzanne Reimer

11 MARDI GRAS AS TOURIST SPECTACLE
Lynda T. Johnston

AGEING AND PLACE

Perspectives, policy, practice

Edited by Gavin J. Andrews and David R. Phillips

LONDON AND NEW YORK

First published 2005
by Routledge
4 Park Square, Milton Park, Abingdon, Oxon OX14 4RN

Simultaneously published in the USA and Canada
by Routledge
605 Third Avenue, New York, NY 10017

Routledge is an imprint of the Taylor & Francis Group, an informa business

Typeset in Baskerville by The Running Head Limited, Cambridge

British Library Cataloguing in Publication Data
A catalogue record for this book is available from the British Library

Library of Congress Cataloging in Publication Data
A catalog record for this book has been requested

ISBN13: 978–0–415–32044–3 (hbk)
ISBN13: 978–0–415–48165–6 (pbk)

CONTENTS

FIGURES AND TABLES

Figures

Tables

CONTRIBUTORS

Gavin J. Andrews was formerly a Reader in Health Studies at Buckinghamshire Chilterns University College and he is currently an Associate Professor at the Faculty of Nursing, University of Toronto, Canada. His research takes a geographical perspective and investigates both the care of older people and nursing policy and practice. Within this rubric, many of his studies have considered the provision and use of private- and voluntary-sector services including residential and nursing homes and small-business complementary medicine. To date, he has published 35 papers in peer-reviewed journals, seven book chapters and five research reports, and is the North American editor of *International Journal of Older People Nursing.*

Helen Bartlett was appointed Foundation Director at the Australasian Centre on Ageing, the University of Queensland, in 2001. Since 1986 she has held research and lecturing positions in social policy and health care in the UK, Western Australia and Hong Kong. From 1995 to 2001 she was Professor of Health Studies and Deputy Head in the School of Health Care at Oxford Brookes University where she also established and was director of the Oxford Centre for Health Care Research and Development. In 1989 she was one of the founding Co-Directors of the Oxford Dementia Centre. Her research has focused on quality and policy issues in community and aged care. She is currently involved in projects on ageing policy, healthy ageing and residential aged care. In addition, she has undertaken numerous evaluation studies of community, health and residential care services/policies for older people for the Department of Health and local authorities in the UK, and for Commonwealth and State Government in Australia. Her publications include three books on nursing homes and continuing care and numerous papers on ageing and aged care in refereed journals. She is on the editorial board of the *Australasian Journal of Ageing*, the *Hong Kong Journal of Gerontology* and the *Journal of Integrated Care.*

Andrew Blaikie is Professor of Historical Sociology and Head of Department of Sociology at the University of Aberdeen. He obtained his MA from the University of Cambridge and Ph.D. from the University of London where, following a spell as research officer in the social history of old age, he was

Lecturer in Gerontology, founding the M.Sc. programme in Life Course Development. In 1991 he moved to Aberdeen. Alongside a continuing interest in ageing and the life course, his research interests have focused upon historical sociology of the family, unmarried motherhood, and, increasingly, photography and collective memory. Among his publications are numerous articles and journal papers and two major monographs: *Illegitimacy, Sex and Society* (Oxford, 1994) and *Ageing and Popular Culture* (Cambridge, 1999). He is currently writing a book entitled *Scottish Lives in Modern Memory*.

Bill Bytheway is Senior Research Fellow in the School of Health and Social Welfare at the Open University and a member of the Centre for Ageing and Biographical Studies. He was editor of *Ageing and Society* between 1997 and 2001 and is author of *Ageism* (Open University Press, 1995). He is currently researching age discrimination and the use of chronological age.

Kevin H. C. Cheng teaches in the Department of Politics and Sociology and has been a research assistant for a multi-disciplinary project on ageing and the environment at the Asia–Pacific Institute of Ageing Studies, Lingnan University, Hong Kong. He received his B.Sc. at the University of New South Wales, MA at the University of Western Sydney, Sydney and his Ph.D. at the University of Hong Kong. His research interests include personality and organisational behaviour, work stress and self-concept. He has published papers in *Applied Psychology: An International Review* and *Ageing International.*

Denise Cloutier-Fisher obtained her Ph.D. from the University of Guelph in 2000. She came to the University of Victoria (UVic) in July 2001 with a joint appointment in the Centre on Aging and the Department of Geography. Prior to coming to UVic she was the National Research System Research Coordinator for the College of Family Physicians of Canada for four years and more recently was a senior population health consultant for three years working with the Central West Health Planning Information Network in Hamilton, Ontario. Her research interests include rural ageing, impacts of restructuring and the challenges of rural health and social service provision. Recent research has focused on the ways in which restructuring has increased the vulnerability of certain elderly and disabled populations. Her research designs emphasise the complementarity of quantitative and qualitative research methods to understand experiences of ageing in rural environments.

Robyn Findlay is a Professional Coach, Psychologist and Director of *ActionToday*, which specialises in empowering talented people to live a balanced and winning life. A former Research Psychologist with the Australasian Centre on Ageing and with the Centre for Accident Research and Road Safety, she began her career as a secondary school teacher. She is a consultant for academic, government and non-government sectors and has a special interest in helping people manage transition periods in their life. She has a Diploma in Teaching, a Bachelor of Human Movement Studies, an Honours degree and a Ph.D. in psychology from the University of Queensland. Robyn has published on a wide range of issues including social isolation and ageing, physical activity and ageing, ageing research and public policy.

Caroline Holland is a Research Associate at the School of Health and Social Welfare, Open University, UK. She has degrees in geography and gerontology. Her interests are centred on environments and ageing and in particular the social, physical and technological aspects of homes, housing and neighbourhoods.

Julia Johnson is Senior Lecturer in the School of Health and Social Welfare at the Open University and a member of the Centre for Ageing and Biographical Studies. She has extensive experience of practice, teaching and research in gerontology. She has co-edited several books on ageing and community care, and is currently a review editor with *Ageing and Society*.

Alun E. Joseph joined the Department of Geography at the University of Guelph in 1978 and served as Chair from 1992 to 2000. Since then he has been the Dean of the College of Social and Applied Human Sciences. His research has focused on rural populations, geographies of health and health care and geographies of care-giving for older people. He was a founding member of the Canadian Aging Research Network and has published more than 70 papers and book chapters.

Robin A. Kearns obtained his Ph.D. at McMaster University (Canada) in 1987 supported by a Commonwealth Scholarship. He is now Associate Professor in the School of Geography and Environmental Science at the University of Auckland. His research centres on the geography of health and health care and the cultural politics of place. He is author of 75 refereed articles and has co-authored two books with Wilbert Gesler, the most recent being *Culture/Place/Health* (Routledge, 2002). Robin is a member of the New Zealand HRC's Public Health Research Committee and serves on the Board of CHRANZ (the Centre for Housing Research, Aotearoa New Zealand). He is on the editorial boards of a number of journals including *Social Science and Medicine*, *Health and Place*, and *Health and Social Care in the Community*. He is currently editor of the *New Zealand Geographer*.

Leonie Kellaher is the Director of the Centre for Environmental and Social Studies in Ageing at London Metropolitan University. She is an established researcher of age and ageing who, over the years, has published a wide range of articles and book chapters. In particular, her research takes sociological and anthropological approaches to material cultures of ageing. Residential homes and communal living have been particular areas of interest for her.

Pia C. Kontos obtained her Ph.D. from the University of Toronto, Department of Public Health Sciences. She is presently a Postdoctoral Fellow at Toronto Rehabilitation Institute, a teaching hospital of the University of Toronto that specialises in adult rehabilitation, long-term and complex continuing care. Her research interests are in critical gerontology, social theories of the body, selfhood and Alzheimer's disease. She has published articles in *Journal of Aging Studies*, *Ageing and Society* and *Philosophy in the Contemporary World*. She has also contributed to several edited volumes, including the forthcoming *Old Age and Agency*, and *Thinking About Dementia: Culture, Loss and the Anthropology of Senility*.

Kevin McCracken, MA Hons (Otago), Ph.D. (Alberta), is Associate Professor in the Department of Human Geography, Macquarie University, Sydney, Australia where he teaches population and medical geography. His research in recent years has focused on geographical dimensions of health, historical population studies, community needs assessment and population ageing. He is the co-author of *Plague in Sydney: The Anatomy of an Epidemic* (1989).

Deirdre McLaughlin is a psychologist who is currently completing a Ph.D. at the University of Queensland, Brisbane, Australia. She has been researching in the field of ageing for the past four years, in particular examining the psychosocial impact of chronic disorders such as epilepsy on the older adult.

Kichu Nair is a Professor of Medicine in the Faculty of Health and Director of Geriatric Medicine at the John Hunter Hospital in Newcastle, New South Wales, Australia. Kichu has published numerous articles both within the field of geriatric medicine and in related areas.

Sheila M. Peace is Sub-Dean, Research, at the School of Health and Social Welfare, Open University, UK. A social geographer and gerontologist, her research interests centre around environments and ageing and in particular care homes and communal living environments. Over the years she has published numerous articles and book chapters and has contributed to the theoretical development of social gerontology in the UK and beyond.

Nancye Peel holds Bachelor of Physiotherapy and Master of Public Health degrees and has had experience in the clinical, administrative and academic sectors. Her research at the School of Population Health, University of Queensland has included health-related quality-of-life issues affecting the older community, particularly in relation to injury prevention. Currently her doctoral research, supported by the Centre for National Research on Disability and Rehabilitation Medicine and the Australasian Centre on Ageing, the University of Queensland, is to examine the determinants of healthy ageing for the development of population-wide interventions to promote healthy ageing as an intervention for prevention of fall-related injury.

David R. Phillips is Chair Professor of Social Policy and was founder Director of the Asia–Pacific Institute for Ageing Studies at Lingnan University, Hong Kong where he is academic Dean of Humanities and Social Sciences. He has broad research interests in social gerontology, with particular foci on epidemiology and ageing, long-term care and housing for older people and population ageing and development in the Asia–Pacific region. In the past ten years he has published a number of books and monographs, including two edited collections, *Health and Development* (Routledge 1994) and *Ageing in the Asia–Pacific Region* (Routledge, 2000). He is co-editor of the *Hong Kong Journal of Gerontology* and an editorial board member of *Social Science and Medicine* and *Ageing International.*

Oi-ling Siu received her Ph.D. from the University of Liverpool and teaches in social psychology in the Department of Politics and Sociology, Lingnan University, Hong Kong. Her research focuses on occupational health and safety, the psychology of ageing and environmental psychology. She is author of articles published in *Psychology of Aging, Accident Analysis and Prevention, Journal of Advanced Nursing, Journal of Environmental Psychology* and *International Journal of Aging and Human Development,* among others.

Janine Wiles completed her Ph.D. in geography at Queens University, Canada and then undertook a one-year Lectureship at St Andrews in Scotland. She is currently an Assistant Professor in the Department of Geography at McGill University, Canada. Her research focuses on social geographies of health and health care. Janine is currently undertaking research on home care in urban communities, exploring the experiences of family and friends caring for a frail, ill or disabled elderly person at home.

Anthony G.-O. Yeh received his Ph.D. from Syracuse University and currently researches urban planning and geographical information systems (GIS) at the Centre of Urban Planning and Environmental Management, the University of Hong Kong. He has published over 20 books and 140 book chapters and articles in international journals, including *Environment and Planning A and B, Urban Studies* and *International Journal of Urban and Regional Research.*

INTRODUCTION

Gavin J. Andrews and David R. Phillips

The focus of this book is on how space and place affect older persons, their families and those involved in services and care. The emphasis is on the diversity of views and the range of relationships and influences space and place have, very broadly defined. These topics are becoming increasingly recognised by policy-makers, planners, service providers and older persons themselves. In particular, it will be evident that ageing and place very much impact on older persons' quality of life and the opportunities for them to achieve goals such as successful and active ageing. The contributors to this book, as outlined below, also adopt very broad perspectives on the topic, and the research themes and disciplines included go well beyond the traditional boundaries of geography. Indeed, mirroring this, the authors come from geography, sociology, environmental psychology, architecture, policy studies, nursing, medicine, public health and social work research. Many of the authors may equally recognise themselves, and be recognised as, social gerontologists who engage in a multi-disciplinary field of study that has grown globally to become the prime forum for social research on ageing. It is quite clear that research on the complex relationships between ageing, space and place is certainly not the preserve of human geography, however broad the discipline may be today. *Ageing and Place* as a concept and as the title for this book therefore naturally and adequately summarises a substantial and a multi-disciplinary field of research in which geographical perspectives do still play an important part.

The book originated from our recognition that although a growing international multi-disciplinary interest in ageing and place had, over the years, been responsible for a rich and varied body of work, there were few dedicated outlets for such research from which it could be easily identified and accessed. After some deliberation, we concluded that two very different publications could be seriously considered. Most ambitiously, a dedicated journal could showcase the continued multi-disciplinary research on ageing and place perhaps to supplement the health-related subjects that have been successfully reported in journals such as *Health and Place* and *Ageing and Society*. However, we felt that an edited collection would initially be useful, dedicated to identifying the major research and policy findings and summarising the perspectives that have evolved to date.

As a book in the series Routledge Studies in Human Geography, this edited collection quite naturally places an emphasis on geographical research and research perspectives on ageing and older people. However, throughout, it also effectively locates geographical gerontology in relation to other disciplines previously mentioned. The book is structured loosely in three sections. These are a set of four introductory, scene-setting chapters, followed by seven thematic chapters considering specific aspects of ageing and place and concluding with three final chapters that critically explore some subtle and very different dynamics between ageing and place. We take this opportunity to introduce briefly the chapters and highlight their interactions with the themes of the book.

In chapter 1, the editors provide an overview of changing issues and perspectives in geographical gerontology and their relationship with broader ageing and place research. Research is traced from earliest books and commentaries and successive reports in *Progress in Human Geography*. Chapter 2, by Robin Kearns and Gavin J. Andrews, effectively takes over from where chapter 1 leaves off and addresses the current state of the art of geographical gerontology and its theoretical 'location' with respect to the parent discipline of geography. The chapter then outlines two recent trends that have both impacted upon, and enriched, geographical gerontology. First, how, following theoretical developments in human geography, health geography has increasingly focused place as being richly constructed by social dynamics and by being a complex symbolic and cultural construction. Second, how other social science disciplines have increasingly become interested in the construction and qualities of places. Kearns and Andrews then explore geographies of ageing and the interdisciplinary 'unification' of a more highly theorised notion of place in three ageing contexts – the body, the home and institutions – and more generally consider the broad concept of 'landscapes of ageing'. An underlying argument is that geographers and gerontologists are now strong producers of ageing and place research and theorisers of ageing environments. A key message is that the research of a greater range of places, and their cultures, could benefit from such perspectives. Indeed, Kearns and Andrews argue that geographical research on ageing has the potential to be a critical strand of human geography.

Chapter 3, by Pia Kontos, complements chapter 2 by providing a critical gerontological perspective. Kontos describes the study of ageing and place as a fast-progressing, progressive and multi-disciplinary research endeavour. Tracing gerontology as a unified science, a humanistic tradition, an anthropological tradition and a critical and cultural tradition, she focuses on how ageing and place have been conceptualised within each. Kontos pays particular attention to the theoretical assumptions made by each tradition, and the methodological approaches used in each to effectively 'unpack' place, as well as highlighting an extensive range of research and research settings to demonstrate her points. The chapter demonstrates how conceptualisations of place have been diverse and the wider 'environment' of ageing research, in which geographers have situated and can situate their enquiry.

In chapter 4 Kevin McCracken and David R. Phillips consider a specific empirical ageing issue, international demographic transitions. Focused on ageing and place at the macro scale, the chapter provides a backdrop to ageing research. McCracken and Phillips discuss population ageing and ask, what regions and countries are ageing demographically? When and how have they aged? They then consider questions that are uppermost in the minds of many policy researchers and practitioners in gerontology, such as what are the consequences are of demographic ageing for population health and more broadly for health policies and health economics? The chapter showcases research at the disciplinary interfaces of population geography, social epidemiology and gerontology. As one of the most established traditions in geographical gerontology, demographic and population research has provided important contextual information, not only for multi-disciplinary research enquiry on ageing but also for governments and health systems worldwide.

Chapter 5, by Gavin J. Andrews, provides a detailed overview of the unique 'semi-institutional' places of residential care homes. In what is truly an area of inter-disciplinary research, Andrews considers certain features and debates with respect to the historical development, location, spatial distributions, regulation and financing of such facilities. Looking inside homes, the chapter then considers design features, ownership and management, caring, residents and home life. Andrews indicates strategic future areas of academic enquiry within this important and well-established area of research. More generally, the chapter outlines the wide range of research which has informed, and to some extent created, debates on the appropriateness of residential care which is in many ways a controversial topic. To some extent it represents institutionalisation; to some extent it is a community-based resource, a provider of social care; and to some extent it is a necessary part of the housing continuum.

Chapter 6, by Janine Wiles, complements the previous chapter and considers home as a site for care provision and consumption. The chapter considers the increased demand for and provision of home care, the structure of home care and recent reforms that have impacted upon it. It discusses the roles of formal and informal care-giving and, in particular, the role of geographical proximity with respect to the caring relationship and home care as a gendered phenomenon. Wiles provides a critical understanding of home care and explores the nature of home and its effects on care, and the nature of care and its effects on home. Both the location and contribution of geographical research is demonstrated for this expanding and important topic of research.

The concept of community living is further developed in chapter 7, by Helen Bartlett and Nancye Peel, who outline a community perspective on healthy ageing. In particular, the chapter explores healthy ageing and its effective measurement. Policy perspectives are also considered and healthy ageing explored through a settings-based approach. Space and place clearly have prime importance to healthy ageing and particular attention is paid to community walking schemes, community-orientated primary care, injury prevention and age-friendly community standards. The chapter draws on many Australian examples and concludes by outlining some research and practice challenges for the future. Bartlett and Peel emphasise that the research of ageing and place should be conducted in places other than homes and institutions, showing how older people are part of communities and neighbourhoods and how they often struggle to maintain their connections with them.

In chapter 8, Kichu Nair considers physical ageing and the use of space. The chapter outlines the relationships between physical frailty and life-space and then engages with a range of issues including, on the micro scale, injury and falls, home design and modification and, on the more macro scale, transport. These environmental design and service issues are revisited in chapter 11, where ageing and urban environments are evaluated. Nair also considers the different uses of space in different residential settings. The chapter provides a practical insight into older people's everyday negotiation of what for some can be a decreasing and sometimes growingly problematic environment.

Chapter 9, by Robyn Findlay and Deirdre McLaughlin, complements chapter 8 by effectively 'placing' the mind, and theoretically exploring environment and the psychological responses to ageing. The authors examine how older people respond via behaviour, feelings and emotions to changes in their environment with a particular focus on adaptive psychological strategies and transactions between psychological, social and physical environments. In particular, the authors critically examine social and ecological theories of ageing, disengagement theories and control and life-course concepts. In doing so, they provide a broad environmental-psychology perspective on ageing. This illustrates the increasingly important influence of environmental psychology in gerontological research and provides a perspective that has developed perhaps the most significant disciplinary engagement with ageing and place.

Ageing within specific communities and locales can highlight major issues in ageing and place. In chapter 10, Alun Joseph and Denise Cloutier-Fisher consider ageing in rural communities, giving particular emphasis to the experience of growing older in restructuring and restructured rural community settings. With reference to research based in Canada and New Zealand and illustrated by examples from in-depth case studies of older individuals, the chapter explores wider social and economic change in rural areas and the problems encountered and the strategies adopted by older people in order to cope with changes in their everyday lives. Again, the important issues of adaptation and modification in respect to environmental challenges for older persons and their families emerge. The complex role of technology as both an enabler and to some extent a creator of vulnerability is also discussed and the authors conclude by considering the transferability of their key arguments to other times and places. Positioned at the disciplinary interface of rural geography and health geography, this chapter demonstrates the complex and constantly changing contexts of ageing in the more isolated areas of the world and how life in them can continue to be a challenge for some older people.

Similar issues are developed in chapter 11, by David R. Phillips, Oi-Ling Siu, Anthony G.-O. Yeh and Kevin H. C. Cheng, who discuss ageing and urban environments. The chapter outlines some potential barriers and difficulties in the design of urban environments with particular reference, at both the macro and micro scales, to accessibility, risks and hazards. More specifically, the chapter considers falls and injuries, stress and coping, basic needs and autonomy, successful design issues and recreation needs. The chapter highlights many challenges created for older persons and their families by aspects of planning and urban design which are commonplace in most cities around the world. The authors emphasises that to address successfully such challenges and to create environments that are friendly for the demographically ageing populations everywhere, serious inter-disciplinary and inter-sectoral planning and policy-making must be undertaken, involving government, NGOs and the private sector.

Chapter 12, by Andrew Blaikie, heads the final three chapters that critically explore some very different dynamics between ageing and place. Blaikie considers 'imagined' landscapes of age – the psychological associations that are made between particular places and ageing. First, the chapter explores the metaphorical landscape of ageing and the way in which spatial metaphors are used as imagery to describe ageing (such as life's journey, going downhill, reaching a crossroads, mapping retirement). Second, in a section entitled 'Projection: looking forward', Blakie considers some stereotyped places associated with older persons, such as seaside resorts and retirement communities. Third, in a section entitled 'Involution: looking backward', he considers the recreation of past worlds of ageing in the media (including 'soap operas', magazines and novels). This perspective, much aligned to critical and cultural geography and taking a critical gerontological perspective on place, demonstrates how ageing identity is framed by place in an imagined (individual and collective) societal project.

Images, stereotypes and place are further developed in chapter 13, by Bill Bytheway and Julia Johnson, who consider the representation of ageing and place in visual displays, specifically photographs and pictures in books, magazines and exhibitions. As a case study, the authors consider pictures from a high-profile gerontological magazine and, in particular, where photographs are taken, what older people are doing in them and implicit ageism and stereotypes. The authors observe that a rather narrow and negative representation of ageing and place is often presented and, importantly, that images of positive and active ageing are rare. Implicitly cross-cutting the interests of cultural geography, art and media studies, Bytheway and Johnson's chapter uncovers the subtle imaging of age and place.

The final chapter, chapter 14, by Sheila Peace, Caroline Holland and Leonie Kellaher, explores place and older people's self-identity. The authors outline and bring together some key debates and theories from gerontologists who are also psychologists, social geographers and anthropologists, including those who articulate ecology and ageing, spatiality in later life, and the possible place attachments that older people make at national, regional and local scales. Based on a current study of older people living in a range of locations and accommodation, the authors extend existing theories and describe place identity as a component of self-identity. They argue that private space exists often as home space, where attachments are made to and through objects (belongings). They consider semi-private space in the form of gardens and landscapes (experienced first-hand and imagined) and demonstrate how public space exists as neighbourhood and community identity. In particular, the authors position home as crucial to older people in terms of meaning, attachment and interaction. Their theory of re-engagement demonstrates how people constantly generate, reinforce and dissolve connections to environment.

This brief outline of the contents indicates that *Ageing and Place* should be very relevant to students and researchers not only across the entire range of human geography but also in many other social science, policy and health fields. As editors, we asked the contributors, for the most part, to concentrate on context and concepts rather than presenting large amounts of data. Rather, if data are cited, they are mostly in the form of short, focused illustrative case studies situated within a broader analysis. In terms of scope, the book ranges broadly from discussions of demographic ageing and epidemiological change to discussions of ageing, metaphor and identity. Similarly, the theoretical scope of the book ranges from structuralist and positivist perspectives to post-structuralist and postmodern analysis. We are aware, however, that most chapters address ageing in the developed world and that issues in many developing countries are not addressed.

As editors, we were very cheered and encouraged in our work by the enthusiasm of our contributors. This in itself reaffirmed our belief in the growing importance of and interest in ageing and place. Without their willing contributions, this collection of chapters could not have been compiled. We also wish to acknowledge the assistance of Ms Dorothy Kok, of the Institute of Humanities and Social Sciences at Lingnan University, who with painstaking diligence checked and edited the reference lists. We also wish to thank the editors of the gerontological and geographical journals who, over the years, have given us the opportunity to develop work in this area. Finally, we thank the international network of scholars from many disciplines who have provided inspiration, advice and constructive criticism.

1

GEOGRAPHICAL STUDIES IN AGEING

Progress and connections to social gerontology

Gavin J. Andrews and David R. Phillips

The study of ageing has traditionally been wide ranging, involving social scientific and health and social care professional disciplines. It is possible to question the extent to which these disciplines have communicated and combined their efforts in true inter-disciplinary fashions. Still, the academic study of ageing has not lacked sustained and wide-ranging attention and has especially benefited in recent years through the multi-disciplinary study of social gerontology (Hooyman and Kiyak, 2002). One consequence is that it is possible to identify a range of disciplinary perspectives on ageing. These may be related to older people's use of, and relationships with, particular housing types or social care and health services. They may also be related to features of older people's social, economic and cultural lives. Cross-cutting both of these disciplinary and subject contexts, older people's relationships with their environments have been a sustained field of research interest.

Most recently, the concern in ageing research for environment, space and place has become even more widespread. Two reasons may be suggested for this. First, this interest may be part of a wider emphasis on place as a central focus of investigation within a range of social science disciplines. Indeed, as part of cultural turns, many social sciences have recently undergone what may be termed 'spatial turns' and have increasingly embraced the importance of space and place and how they may impact on, and represent, human experiences, behaviour and activity. Moreover, place has increasingly been conceptualised not only in a physical sense, but as a complex symbolic and cultural construction. Second, and more practically, academic interest in space and place has also been motivated by unprecedented demographic, social, fiscal and technological changes that have impacted simultaneously in many countries (McKeever and Coyte, 1999). Indeed, these are well documented, and include rapidly ageing populations, changing kinship relationships and responsibilities, an ever broader range of health and social care and increasingly limited resources with which to provide it. Together they have radically altered, and broadened, both the ways and the places in which health and social care is provided. Importantly, change has meant that the health care system is no longer hospital-based, discrete and bounded, but diffuse. Health care sites now include virtually every setting where human beings reside in, frequent and, importantly, live in (McKeever and Coyte, 1999). Being substantial recipients of welfare and care services, and a social group particularly vulnerable in economic and social change processes, older people have often experienced these changes most acutely. Researchers of older people from a range of disciplines have therefore, almost by implication, found themselves investigating, comparing and conceptualising a wide range of social care, health care and residential settings.

Early beginnings

Just as the study of ageing is multi-disciplinary, so too is the study of ageing and place. Indeed, studies which specifically explore space and place and bring them to the fore in their analysis have originated not only from the more obvious academic disciplines of social and health geography, environmental psychology and architecture, but also from a range of other social science disciplines including sociology, social policy and health economics, together with professional disciplines including public health, nursing, occupational therapy and social work research. In this chapter we briefly outline changing geographical perspectives, but with a view to locating them within the wider multi-disciplinary endeavour of ageing and place research.

The early origins of geographical gerontology can be traced through a small number of key studies and commentaries. The first substantive geographical study on ageing was published over 30 years ago. *The Residential Location and Spatial Behavior of the Elderly* (Golant, 1972) focused on the spatial behaviour of older people in Toronto, Canada, and on local demographic trends and population distributions, local transport and housing issues and activity levels. Golant's later work *A Place to Grow Old: The meaning of environment in old age* (Golant, 1984), bridged early geographical, ecological and psychological disciplinary perspectives on older age, and certainly set the agenda for geographical gerontology to be a sub-discipline highly connected to others. Particular subsections focused on the behavoural relationships between older people and their environments, territorial environment experiences, physical environment experiences and social environment experiences. Certainly the best-known and most frequently cited early work in geographical gerontology is *Prisoners of Space? Exploring the geographical experience of older people* (Rowles, 1978). Notably, this is also a very early example of geographical gerontology to use qualitative research methods. The research described in depth the lives of five older persons living in a working-class neighbourhood of a North American city and investigated their very different responses to growing old over a two-year period and how this is connected with their local environmental context. Engaging critically with other disciplines, the book was the first to argue that older persons' environmental experiences were far more complex than the progressive spatial restrictions described by psychologists in classical disengagement theory and its derivatives (see chapter 9). Five years later, Graham Rowles co-edited another influential book with Russell Ohta, *Aging and Milieu: environmental perspectives on growing old* (Rowles and Ohta, 1983). Similarly ahead of its time, it introduced many new issues to ageing research, including foci on clinical and therapeutic environments; the relationship between time, space and social activity; the meaning of community and neighbourhood; experiences of ageing in rural environments; environmental learning and adaptation; the meaning of place in old age; sex roles and social environment; and the role of place in age stereotyping.

Geographical gerontology began to develop as a distinct and recognisable sub-discipline in the early 1980s and at about the same time as *Geographical Perspectives on the Elderly* was edited by Tony Warnes (Warnes, 1982). This collection was the first to bring together applied geographical research based in a number of countries. With foci predominantly on the distributive features of older population and services for the older population in space, the book very much reflected the theoretical and methodological priorities of the parent discipline of human geography at the time. Particular chapters focused on issues including population movement, retirement migration, specialised housing, activity patterns, travel difficulties and the distribution and planning of service provision. Over the past 20 years Tony Warnes has continued to make many other influential contributions to, and pioneer, geographical gerontology. In particular, his later edited book *Human Ageing and Later Life: Multidisciplinary perspectives* (Warnes, 1989) brought together a range of perspectives from health psychology, medical and social geography and social gerontology. Moreover, Tony Warnes' positions outside geography as head of a gerontology research centre and as editor of the journal *Ageing and Society* have helped to make geographical gerontology visible to a wider academic audience.

The early 1980s also witnessed the commencement of dedicated progress reports and agenda-setting papers in geographical gerontology published in the journal *Progress in Human Geography*. Rowles (1986) reported that geographical research on ageing had flourished but had, to that time, been relatively descriptive in terms of concentrations of ageing population, migration and service delivery. He explained this as an information accumulation phase that precedes the emergence of a dominant paradigm in any new field of research or sub-discipline. Attempting to set a cohesive and comprehensive agenda for geographical gerontology, Rowles' focus was on ageing from a transactional perspective, instead of describing older people as a group, emphasising the relationship between the older person (individual) and their environment. Rowles both reviewed and called for research on the meaning of place and place memories and, in particular, on home. Finally, he called for a greater appreciation of the diversity of environmental contexts in which older people reside and for a greater appreciation of different ages of older persons and different cultures of older persons. The types of place and context included retirement communities, neighbourhoods, retirement homes and migration locations. Notably, he commented that gerontologists had long been taking such perspectives, albeit without explicit geographical foci.

Shortly after, although not necessarily conflicting in his view, Warnes (1990) took a different perspective by reviewing the contribution of geographical gerontology to gerontology and developing an agenda which could maximise this contribution. From the outset, Warnes described this as an uneven contribution and argued that geographers needed to shift their priorities and objectives away from servicing human geography and onto social gerontology and the needs of older people. For Warnes, the most important objective of social gerontology is, in the face of public misunderstanding and stereotyping, to increase human understanding of the ageing process. Warnes highlighted three issues which had not been given enough attention by geographers: the global evolution in demographic ageing and its implications, locational dimensions in the circumstances of older people's lives and temporal change in the interaction between the environment and older people.

The latest progress report on geographical studies in ageing was published almost one decade ago (Harper and Laws, 1995). While Harper and Laws' report makes it clear that some of the issues had been addressed within these broad areas in the preceding five years, it reaffirmed that geographical gerontology remains low-profile and a limited research endeavour. Furthermore, counter and in contrast to Warnes, the authors argued for geographical gerontology to learn theoretical and methodological lessons from human geography and that this would open new possibilities for the geographical study of ageing. The authors went further and argued that these may provide approaches that even gerontologists (non-geographers) might find useful. Harper and Laws critiqued geographical gerontology for continued empiricism, for the dominance of positivistic approaches and for using a general and unproblematic notion of 'old', and for not exploring the causes and nature of age segregation. The authors argued for the greater use of social theory, as has been used by cultural geography, in particular in the areas of political economy, gender roles, sterotypes and the role of the state in service provision, and argued for attention to postmodern life courses through postmodern research perspectives. More generally, regardless of whether geography or gerontology is the more 'critical', 'radical' or 'postmodern', the differences in emphasis between Warnes' commentary and that of Harper and Laws reflects a wider tension and predicament for many geographical gerontologists regarding what discipline to speak from and to – to be a geographer who studies older people or a gerontologist who takes a geographical perspective.

Current perspectives

Almost ten years on from the last review published in *Progress in Human Geography*, we can briefly reflect on current perspectives and theory in geographical gerontology. Following one of the more established research traditions, research has continued to consider demographic transitions and spatial distributions of older people within international and national contexts (Bartlett and Phillips, 1995; 2000). Migration patterns of older people have continued to be an important focus and the specific socio-spatial trend of 'retirement migration' has also been researched more expansively to consider migration across national borders (King *et al.*, 2000). In this respect, research had answered both Rowles' and Warnes' more traditional requests. However, research has been more expansive and spatial distributions of older populations have also been contextualised in terms of changes in family interactions and cohesiveness over time, and the increased spatial separation of family members (Lin and Rogerson, 1995; Greenwell and Bengston, 1997). Another established research tradition again answers Rowles' and Warnes' requests and continues to examine the spatial allocation and use of resources provided by both the private and public sectors, specifically for older people. In particular, the distribution of home-based social care facilities and residential accommodation on the national (Smith and Ford, 1998) and regional scales have been the main foci of this research. These perspectives however have been refined and updated, and the distribution of social care provision for older people has been contextualised in terms of market regulation (Bartlett and Phillips, 1996) and the wider political economy of health and social care (Joseph and Chalmers, 1999; Andrews and Phillips, 2000; 2002). Older people's own experiences of residential settings have also demanded attention, and certain studies provide insights into residential home life (Higgs *et al.*, 1998). Consistent with an international policy emphasis on community-based care, are an emerging number of studies which focus on older people's own homes as new sites of health care consumption (Wiles, 2003a; 2003b) and other support services (Andrews *et al.*, 2003). Taking a novel approach, uses of complementary and alternative medicine have also become an interest and how older people have gained empowerment and health benefit through what are heavily emplaced experiences in the community and home (Andrews, 2002a; 2003).

Another strand of research is concerned with older persons' personal and evolving relationships with places and spaces. Much of this more recent research has considered the environmental constraints affecting older people's access to services (Joseph and Hallman, 1998; Stainer, 1999). Also investigated are the problems encountered by older people in both urban and rural living environments, and the consequences for local planning and wider social policy (Bartlett and Phillips, 1997a; Cloutier-Fisher and Joseph, 2000). Increasingly attention is paid to the immediate environmental components that affect older persons' abilities to achieve successful ageing and in the context of constraints (personal and environmental) (Phillips *et al.*, 2004). Indeed, internal environments in the home are, for example, increasingly recognised as imposing constraints and often offering hazards and allowing accidents to occur (Carter *et al.*, 1997). More generally, environment and ageing are now key planning issues in many rapidly ageing countries, for example in the Asia–Pacific region, where interest extends to interactions with long-term care, housing and other areas of social policy and service provision (Phillips, 2000; Phillips and Chan, 2002). The planning aspects of urban expansion, urban renewal and redevelopment have also become key foci in such areas (Phillips and Yeh, 1999). Meanwhile, the concept of ageing-in-place has been analysed as a complex geographical process (Cutchin, 2003).

These above areas of emerging research have started to address Rowles' early concerns for emphasis in research to be placed on the meaning of place and on individuals. However, despite these developments, a pertinent question is whether or not research has addressed Harper and Laws' more recent requests for postmodern perspectives, the integration of social theory and perspectives from cultural geography. If anything, however, despite a small number of exceptions (for example, Cutchin, 2003; Wiles, 2003a; 2003b), during the past decade social gerontologists have been most active in integrating sophisticated theory in ageing and place research (for example, Kontos, 1998). This has been undertaken quite independently from mainstream debate in cultural geography. Indeed, research considers the way that various forms of media are involved in imaging of ageing and place (Blaikie, 1995; 1997; Featherstone and Wernick, 1995), how particular sites of older people's care hold strong cultural and symbolic meanings for older people (Hockey *et al.*, 2001), inform societal attitudes of older age (Gubrium and Holstein, 1999) and are associated with negative images and stigma (Bytheway, 2003). Indeed, ironically, Harper and Laws' requests have not been fully explored by geographers and an important question for the future remains the potential for geographers to follow the lead of social gerontologists and develop their own postmodern and critical geographical gerontology. Given the emergence of critical perspectives and social theory in geographies of care, caring and the body (Parr, 2002; 2003) and in geographies of disability (Butler and Parr, 1999), the potential for such has clearly been signposted elsewhere in geography.

At the time of writing, the Tenth International Symposium in health geography had recently been held in July 2003 at Manchester, UK. Notably, Mark Rosenberg's opening keynote address called for health geographers to take age and ageing much more seriously in their research. Rosenberg's calls were based on a simple observation: that although some innovative research on ageing had occurred in health geography, the volume and range of work does not reflect the varied challenges that ageing populations pose to health and welfare services worldwide. His comments were verified at the conference in that although many later papers connected with ageing indirectly (older people being a proportion of service users or patients) only three of over 100 papers dealt directly and specifically with ageing as a distinct subject. To speculate, one might expect to find the same lack of attention to ageing in social and cultural geography conferences and significantly more attention to space, place and settings at social gerontology conferences. Arguably, then, although much progress has been made, many challenges remain for the future, both in terms of geographical issues to be studied and theories that are applied and developed to study them. This book indicates the range of issues and perspectives at the heart of such a project.

2

PLACING AGEING

Positionings in the study of older people

Robin A. Kearns and Gavin J. Andrews

Ageing is not only an embodied process but is emplaced as well.
(McHugh, 2003: 169)

Introduction

Two trends have signalled an integration of research on ageing and research on place within recent scholarship: a placement of geographical work within the broader field of health geography, and the embracing of the fundamentally geographical theme of place by researchers in other disciplines. This chapter selectively reviews the recent shifts in thinking among geographers that have – at least in name – transformed medical geography and brought with them new possibilities for incorporating concerns with ageing into geographical scholarship. Such shifts led to a reconceptualised 'health geography' and have involved a philosophical and empirical distancing from concerns with disease and the interests of the medical world. The net result has been the deployment of a sharper focus on people's situated well-being often invoking social models of health and health care. These moves within geography over the last 15 years have occurred contemporaneously with a key shift beyond geography, as suggested: the increasing centrality of place within research elsewhere in the social sciences. Given these contexts, this chapter reviews ways in which research on ageing and place can be interpreted through the lens of a medical geography, recently revitalised through a health orientation. First, the centrality of place in health geography is considered. Second, three key sites in the geography of ageing are examined: the body, the home and the institutional setting. Third, the application of the more generalised concept of 'landscapes of ageing' is considered. A short set of conclusions closes the chapter.

The centrality of place within health geography and ageing research

Just as birth gradually became the domain of doctors and hospitals through the twentieth century, so too the daily experience of ageing, and changing health status in older age, have tended to become medicalised. Reflecting a broader problematising of older age within society, much research on geographical aspects of ageing has conventionally been 'placed' within medical geography. Indeed. the sub-discipline of medical geography has traditionally been divided according to health status and service characteristics and most geographical research on older age has fallen under the latter category. Through the 1990s, however, a variety of critical research perspectives, involving political–philosophical standpoints (e.g. feminism) and methodological approaches (i.e. the qualitative 'turn'), highlighted an over-emphasis in research on disease and disease services. The resulting move towards a 'post-medical geography of health' (Kearns, 1993) was an attempt to be informed by both political and post-structural perspectives and move beyond the shadow of medicine. These shifts in thinking have transformed medical geography and have led to a reconceptualised 'health geography', which has incorporated research on health and wellness. Advocacy of a self-consciously health geography was also associated with novel conceptualisations such as 'therapeutic landscapes' (Gesler, 1992), which allowed an emphasis on the power of place, and how it shapes health beliefs and invokes situated experiences of well-being. Within a reconstructed health geography, these trends have returned focus to the person, their body and their lived situations rather dwelling on disembodied diseases. As will be shown, research on ageing and its place, from across the social science disciplines, has both anticipated and reflected these shifts.

Much of the work undertaken by health geographers is informed by the so-called 'cultural turn' within the social sciences. Researchers have been influenced in their work by postmodernism, post-structuralism and social constructivist perspectives that have resulted in new research directions and challenged knowledges previously taken for granted. These shifts are particularly evident in the changing conceptualisation of 'place' within research. Under the rubric of medical geography, space and place were treated as unproblematic, either as a series of geometric points and boundaries concerned with distance or location, or as a passive container in which characteristics are recorded. Health geographers have argued, however, that place 'matters', and are increasingly concerned with the meaning, experience and construction of place and how these influence health. Place has been operationalised as a 'living construct' rather than as a blank slate upon which events are inscribed. In Kearns and Moon's recent (2002) review of the positioning of health geography within human geography and health-related research, three significant themes were identified: the centrality of place within geographical concern for health, the placing of research within self-consciously socio-cultural theoretical frameworks, and the quest to develop critical geographies of health. With these themes in mind, this chapter reflects on the way in which 'place' as a central organising concept has been incorporated in research on older people. It will also be suggested that the themes at the very heart of health geography offer much potential for repositioning enquiry into ageing and place. An important point to remember, however, is that, as suggested, many studies in ageing and place have originated, and may well originate in future, from outside the sub-discipline of health geography, for example from sociology, cultural studies and health services research. Thus matters of place are by no means the sole domain of geography but are increasingly inter-disciplinary in nature, reflecting the extent to which the cultural turn that has permeated the social sciences includes a spatial turn. The reinvigorated concern with place identified by Kearns and Moon (2002) can be discerned in four domains of work on ageing that extend in an outward scale from the person: the body, the home, residential care settings and more generalised landscapes of ageing.

The body

A key exhortation by early advocates of a reformed medical geography was greater attention to the place of the body (Dorn and Laws, 1994). As Kearns and Moon (2002) commented, medical geographers have since begun to address geographies of the body, their research foci predominantly being on psychologically or physically diseased, impaired, marginalised and disadvantaged bodies. Parr (2002) supported this observation, and noted that geographical research on the body provides a critical perspective to medical places considering the body in wards, clinics and community settings; the body in relation to diagnosis and medical power; 'othered' bodies; and historical geographies of the ill body and mind.

Within the broader expansion of research interest in embodiment in the social sciences, the leading questions have become not only how but 'where does the body as a corporeal presence serve as a surface for the assignment of personal meaning and an organising principle for social interaction'? (Gubrium and Holstein, 1999: 520). Hence the body is both a geography unto itself and is assigned meaning through the geographies in which it dwells. The nursing home, for instance, is a common institutional basis for everyday life for many older people. Here, the body becomes a surface of signs, monitored for evidence of varied concern for stakeholders – from the resident's own maintenance of identity to family members' 'lingering sense of responsibility after placement' (Gubrium and Holstein, 1999: 520). In distinction to the hospital (where the discursive anchor – perversely even in the face of terminal illness – is recovery), a nursing home is discursively lodged within decline. In other words, even against the odds, people are expected to graduate from hospital into greater independence, but the reverse prevails in the nursing home: it would be counter-intuitive to graduate into anything but greater dependence (e.g. a geriatric ward) (Gubrium and Holstein, 1999).

Older age may be imbued with positive connections to past events, traditions and histories, all of which have strong connections to places (Hugman, 1999). However, more frequently, older age is closely associated with physical and mental deterioration and decline of the body (for example changes in skin tone, body shape, body structure, mental capacity) and a range of places which are purposefully designed and provided to cope with such limitations. In this context, nursing and residential homes represent, both literally and in terms of image, concealment and disengagement of the body from society. In contrast, other places, such as retirement communites and home, represent the body more positively in terms of identity, independence and ability (Hugman, 1999).

These negative body and place images may be over-represented. For example, even though only a small proportion of all people develop dementia, corresponding places of 'containment' (such as nursing homes) receive most public and academic debate, making the able majority and their living spaces somewhat invisible (Hugman, 1999). Here lie the foundations for societal stereotyping and even discrimination based on a narrow version of ageing. Furthermore, this is supported by a societal emphasis on youth as positive and 'the norm'. In short, neither research nor society has sufficiently addressed the normality of ageing, the whole person and the range of places frequented. One important direction for research therefore is to also study the ageing body in positive contexts such as sexuality, fitness, spirituality and consumerism and to conceptualise their relationship to places.

The home space

The majority of research on ageing, health and place has focused on the experience, meaning and construction of two locales: the home and supportive residential care facilities. Much research focusing on these two sites contributes to the enduring debate within social gerontology of the relative merits of each setting in terms of well-being (Cutchin, 2003). Undoubtedly, however, the home space has considerable influence on the health and well-being of all people as it is where we spend much of our free time. It is especially important for older people, some of whom may experience diminished mobility and may be unable to regularly venture outside of their home. A number of studies have considered the experiential aspects of the home space with attendant implications for health. For instance, Twigg's (1999) exploration of the spatial dimensions of bathing support services for older and disabled persons within the home employs the notion of private versus public space. With the advent of health service provision 'in the community', the home is arguably now the most important site for the provision of community services. Twigg argues for a number of spatial dimensions along the private–public continuum that are important in understanding the home space, such as the way in which 'home' can bestow on the occupier the ability to exclude and include. The home space is also a locus of control and order, aspects which are considered to influence health and well-being. Not only is the home space structured within the private versus public space dynamic, but spatial ordering also occurs within the home itself. Twigg suggests there is a privacy gradient, where halls are more public than sitting rooms, which, in turn, are more public than bedrooms. Disability, infirmity and the need for care, however, reorders the spatial organisation structures within the home as the individual may be restricted to a single space. On a related note, Percival (2002) considers the micro-geographies of home, addressing the way that older people use, perceive and attribute meaning to spaces within the home through everyday activities such as entertaining, housework, leisure and decoration. He argues that domestic space is particularly important for the establishment and maintenance of personal identity, control and, ultimately, well-being. At the same time, however, some older people with limited mobility report feeling trapped within the confines of their home, a theme explored by geographer Graham Rowles in his pathbreaking book *Prisoners of Space?* in 1978. While Rowles' study refuted the notion that older people's worlds were small, Percival highlights the challenges of scale when lived environments are small-scale. Participants in his study criticised the size of their homes and the accessibility and maneuverability issues this poses. One respondent criticised the way in which designers assume that older people do not require much space, an assumption that is perpetuated within the built environment. In the words of one participant, 'So far as they're [housing providers] concerned, you don't need a lot of space. You are disabled so you should sit in the chair and that's it. You won't need space for moving around' (Percival, 2002: 740).

Such a narrative speaks to the difficulties of navigating around, and living within, a restricted space and how this can erode a sense of personal control and self-determination. However, if a restricted geography of everyday life means a loss of control, this can be compensated by adornment and display. Just as gardens and houses can be media for self-expression and pride of place among the young and fully able-bodied, older people have been shown to personalise their homes through decoration or display of valued objects. It is suggested that this may be to enable reflection, provide memory prompts and imbue their home space with personal meaning (Percival, 2002). Ultimately, however, the home space seems to be regarded with some ambivalence. Mowl *et al.* (2000) found that for some elderly respondents, exclusively spending time in the home space was identified as health-damaging, accelerating decline and ultimately leading to death. For others, being able to still reside in their own home was seen as a symbol of resistance to the ageing process itself.

Recently, a key transition in research is from thinking of place as location or container for older people, and moving through a more theorised engagement with the politics of location. Moss (1997), for instance, invokes ideas from political economy to offer a socio-spatial analysis of the ways older women with disabilities (rheumatoid arthritis) navigate and (re)structure their home environments. Her case studies illuminate the ways in which women adapt their home environments (physical and social), including structural modifications to the home, such as bathroom railings. Critical geographies of ageing and place cannot exclusively dwell with the individual or within the home. In response, Moss also considers the influence of wider political economic structures by addressing the women's access to external resources. On a related note, Wiles (2003) considers the relationship between older people and informal caregivers in terms of routine and scales located both inside and outside of homes.

Residential care facilities (RCFs)

A locale that has received considerable research attention is that of the residential care facility (RCF). A great deal of this research is reflective of medical geography's tradition of focusing on distributive features of health services and hence discusses access to, and the siting of, RCFs. However, increasingly, from both inside and outside geography, studies seek to explore the nuances of older people's experiences of, and meanings associated with, these places and their implications for well-being. Studies have both supported and opposed RCFs in what has become known as the 'residential care versus community care debate' (Andrews and Phillips, 2002). The critique of RCFs has generally been twofold. The primary strand of criticism has focused on standards, the inability of RCFs to provide independence and choice for residents and to protect their civil liberties, and the inability of regulating bodies to ensure adequate quality standards (Andrews and Phillips, 2002). For example, Peace *et al.* (1997) argue that, despite greater regulation, RCFs continue to depersonalise older people and Twigg (1997) argues that carers are more likely to be controlling and obtrusive in residential settings. Recently Fisk (1999) suggested that residential care has conceptually reached 'the end of the line'. RCFs are criticised for reflecting institutional patterns of provision that borrow from historical Poor Law philosophies, concerned with custody as much as with care (Andrews and Phillips, 2002). Another strand of criticism views homes as places of stigma and social marginalisation. From this perspective, no matter how much standards may be improved, residential homes will always have negative associations that further marginalise and negatively categorise older people (Andrews and Phillips, 2002).

Alternative views on RCFs are, however, emerging. Oldman and Quilgars (1999) comment that the 'in home/at home' literature is dominated by a structured dependency paradigm which depicts residential care (in a home) as institutional and 'at home' as the only means to personal freedom, and note that many authors have started to question assumptions implicit in critiques of residential care (Andrews and Phillips; 2002). Their own empirical research found that some older people experienced their home not as a place where they found self-expression and autonomy, but as a lonely and bleak place, while some residents of RCFs enjoyed the lifestyle and independence of the setting and took comfort in that they were not being a burden on relatives. Baldwin *et al.* (1993) argue that researchers often ignore the quality of older people's lives prior to moving into residential care. These authors posit that residential care cannot be blamed for generating dependency and a gradual reliance on others. As Andrews and Phillips (2002) have suggested, in their view, 'institutionalisation' may occur regardless of the care setting. Reed and Royskell-Payton (1996) also challenge the stereotyping of residential care and suggest that the residents of homes are not institutionalised victims separated from community and society, while Noro and Aro (1997) found that the majority of residents in homes would prefer to stay in residential care, rather than be cared for at home. We can only conclude from this debate that a diversity of experiences among residents of RCFs exists, as does the need for fine-grained analyses and understandings of ageing and place. Despite the foregoing issues being a source of long-standing argument in political and academic communities, this essentially geographical debate has paradoxically been engaged only to a limited extent by geographers, who have, by default, deferred to social gerontologists (Andrews and Phillips, 2002).

Solutions to the problems of RCF may lie in new types of service. Indeed, in this context research by Biggs *et al.* (2000) explores narratives of those living in retirement communities, which are argued to contain the best elements of both RCFs and home. The authors employed a multi-layered approach examining four dimensions of settings: the way that the communities are represented in printed promotional material; the experiences of residents who dwell within the community; the influence the RCF has on well-being; and the metaphors, meanings and themes that connect both representation and experience of retirement communities. Residents stated that the community was a place where they could be independent yet which offered security and inclusion. In other words, it combined the independence of living in one's own home within the wider community with the safety and security of a formal RCF. Interestingly, nursing homes (in dependency terms, the next care category 'up' from RCF) were disdainfully spoken of – as 'unhealthy' places that inhibit independence and autonomy. Residents associated community settings with promoting higher levels of health and well-being through the culture of the place, emphasising the strength in the will of individuals in determining both the cause and resolution of ill health. Residents perceived themselves to be part of a community which, rather than focusing on illness and degeneration, focuses on ability, mutual interdependence and well-being. However, notably, Biggs *et al.* also pointed out that there were some residents who felt excluded from the dominant narratives of belonging and well-being.

The preceding discussion has considered the way in which place-related research on age and health has tended to focus on the home and the RCF. In addition, there are a few studies that consider movement between the two. For example, Cartier (2003) considers the relationship between institutional health care spaces and the experience of receiving health care services at home. This relationship is framed within the context of economic restructuring of health services which, Cartier argues, has influenced how people experience the places in which they receive care. The author argues that the places in which health care services are provided are important for a variety of actors (i.e. those receiving and delivering care in addition to family, friends and health professionals). Cartier argues that the fragmentation and diffusion of health care services has occurred within discourses of economic savings and 'community is best' care, yet it is suggested that this cost saving strategy has simply transferred costs from institutions to the individual and their family. On a practical level, there is evidence to suggest that shifts between sites of care (particularly in light of policies governing how long a patient can stay at hospital), can be health-damaging. On a related note, Reed *et al.* (2003) comment that too easily older people can be objectified and unintentionally rendered passive in research on RCFs. In this context, the authors considered the extent to which older people themselves play an active role in relocation decisions. They argue that

> we need to think about older people and their relationship to the care-home in a different light. Rather than conceptualising older people as to be 'placed' in a care-home, which implies passivity, we should think about them as 'living' in the home, where they make active choices, decisions and judgements about their living environment.
>
> (Reed *et al.*, 2003: 239)

A later chapter in this book will consider the multi-disciplinary study of RCFs in much greater detail and the part that geography plays in it. However, from this short section it is evident that, even if not explicitly geographical in nature, research has started to unravel the finer features that make these places in terms of civil rights, independence, privacy, everyday home life, image and identity.

Landscapes of ageing

'Landscape' is a key construct within contemporary cultural geography that has found ready application within health geography. A range of places have been identified as revealing evidence of long-standing social processes and physical endowment that either enhance or detract from well-being (e.g. Dear and Wolch, 1987; Gesler, 1992). Some geographers have explicitly applied this construct to themes related to ageing (e.g. Laws, 1993) and others have simply written as if informed by its geographical sensibility. Blaikie (1997), for instance, links ideas of positive ageing with close-knit coastal settlements, highlighting the ways that images of ageing are used in promoting retirement in coastal areas. Beliefs about the moral and spiritually uplifting values of coastal communities are thus mapped onto aged bodies in a way reminiscent of the 'therapeutic landscape' concept adopted and adapted by health geographers (Williams, 1999). In this sense the coastal landscape is 'therapeutic' not just in terms of its physical or climatic character, but also through its imputed social characteristics, and historical layerings as a healthy place (see Andrews and Kearns, 2004). These connections lead Blaikie to suggest that 'older people are perhaps agents in the reconstruction of communities in their own desired image' (1997: 630).

New landscapes also exist in terms of the character of health and social care provision. Indeed, as population structure has shifted towards more aged distributions in developed countries, private market and public policy responses have invariably interacted in creating new social geographies of ageing (Andrews and Phillips, 2002). There have been a number of applied studies of the spatial outcomes of such processes (e.g. Phillips and Vincent, 1986; Smith and Ford, 1998), as well as the characteristics of provision (Vincent, Tibbenham and Phillips, 1986; Andrews and Kendall, 2000). McHugh (2003), for instance, critiques images and texts produced about 'successful ageing' as 'anti-ageing' and the spaces and places inherent in these representations, documenting the competition between various locations in the US in attracting older people. However, although geographers have explored the cultures of patients and recipients of care with respect to mental illness (e.g. Philo, 2000; Pinfold, 2000), the same attention has not been devoted to elderly persons since the pioneering work of Rowles (1978). However, Andrews and Phillips (2002) signal a new direction in their exploration of local cultures of provision.

A deeper engagement with the processes producing landscape has been prompted by the imperatives of recent restructuring. Following the metaphor of 'changing places' used by Le Heron and Pawson (1996), geographers have noted the ways that actual places and metaphorical place-in-the-world have been recursively reshaped through economic and social policy restructuring processes. Andrews and Phillips (2002), for instance, draw on Kearns and Joseph's (1993) advocacy of a more self-conscious consideration of structure and agency in health geography in their consideration of private residential care for older people. Their linking of macro-level policy shifts to micro-level agent changes is echoed in an earlier examination of long-term care restructuring in rural Ontario by Cloutier-Fisher and Joseph (2000). Given contemporary economic and service-sector restructuring, they advocate developing a 'situated understanding' of the changing place of care (both literal and metaphorical) through considering both service-user and service-provider perspectives. Such analysis builds on recognition that parallel geographies exist: the publicly funded care system, and a 'shadow system' of informal care provided by friends, family, neighbours and local and less visible local institutions (Joseph and Martin-Matthews, 1993). Cloutier-Fisher and Joseph's (2000) approach has a methodological perspective that is in keeping with the general trend towards the qualitative turn in health geography: they advocate collecting and analysing narratives from users and providers of services in order to allow 'shading in the middle ground of understanding' (2000: 1038).

In future, ageing and place may be examined using the therapeutic landscapes concept. This approach allows an emphasis on well-being in place and is informed by cultural–geographic notions that landscapes can be 'read' (Duncan, 1990; Duncan and Duncan, 1988). The idea of therapeutic landscapes involves certain places having reputations for, and subsequently being perceived to have, therapeutic value (Gesler, 1992). Some pointers for future research are apparent in existing published research. As suggested above, within the context of ageing, while sociologists such as Blaikie (1997) do not explicitly use the term therapeutic landscape, they have nevertheless investigated places that are commonly associated with health and, in this case, healthy ageing. Blaikie's analysis of visual images, for instance, allows an illumination of the association between images of seaside towns and positive, healthy ageing. He suggests that coastal towns become categorised as virtuous havens from the sins and ills of the city back in Victorian times. This is akin to Gesler's (1992) notion of therapeutic landscapes whereby seaside towns are infused with health-promoting qualities (Andrews and Kearns, 2004). Shifting from the natural to the urban, Williams (2002) examines the home space as a therapeutic landscape for care-givers and recipients, a space which has become increasingly important in the provision of health and well-being services, particularly after deinstitutionalisation and a shift to community-provided services. Williams suggests that a good example that demonstrates health restructuring is the case of older people, where the burden of care has been transferred from institutions to families. Williams particularly focuses on one facet of the notion of therapeutic landscapes: the way in which landscapes can maintain health and well-being (as opposed to the more widely known dimension which emphasises the use of therapeutic landscapes for healing and recovery from illness (Gesler, 1992)). Elsewhere, Andrews seeks to examine the interconnections between consumption of complementary medicine (CM), therapy and place. Andrews argues that for older users a strong 'sense of place' can be observed in the consumption of CM. Not only were the CM clinics important to the overall experience, but also the towns or villages in which they are located. Andrews provides a case study of Totnes, a town in Devon, England known for its embrace of alternative lifestyles. Andrews points out that users reported that treatment improved their mobility and enabled a wider use of space and the physical environment. Users reported the therapeutic value of the treatment session but also the application of the practitioner's advice in their home space. Therefore, 'Older users potentially controlled and extended their therapeutic experiences to incorporate their everyday social lives' (Andrews, 2003a: 8).

Conclusion

This review has been necessarily selective. Indeed, any review of geographical concerns with ageing must be qualified with an acknowledgement of the various range of scales in which it can be undertaken. For example, reports on the rates of mortality and disease among older people range from macro-level comparisons in European countries through to analysis of the state of the health of the elderly in small settlements and the inner city (Gazerro *et al.*, 1996; Warnes, 1999). However, a pertinent question regards what unifies new positionings in ageing and place research. The answer must be creativity of method and an expansion of theory. Place encourages, is formed by and incorporates talk. As Gubrium and Holstein (1999: 521) argue, the very term 'nursing home', for instance, 'guides and ramifies body talk relevant to disease, caregiving and dying'. Talk about place can thus constitute data, as the narratives collected shade in the 'middle ground' between people and place (Cloutier-Fisher and Joseph, 2000). In another sense, talk becomes galvanised in commonly held discursive images of place associated with the 'land of old age' (Laws, 1993); the coastal images analysed by Blaikie (1997) are, in a sense, a visual 'talk' framing experience and putting older people into the picture.

Just as social models of health emphasise well-being over the life course rather than specific, and medicalised, events, so too has medical geography been reinvigorated by concern with the in-place well-being of age-defined population groups. In being re-articulated as health geography, the sub-discipline has incorporated the concerns of social and cultural geography and hence moved closer to the heart of the discipline of geography as a whole. Yet, contemporaneously, concerns with health have led to a catholicity of outlook among geographers interested in ageing that is inherently trans-disciplinary in nature.

Acknowledgement

The authors would like to thank Jody Lawrence for her assistance in preparing this chapter.

3

MULTI-DISCIPLINARY CONFIGURATIONS IN GERONTOLOGY

Pia C. Kontos

Introduction

Old age, the inevitable encounter with the frontier of life and death, is a multi-dimensional phenomenon. It embraces notions of life history, continuity and change, activity and disengagement, autonomy and dependency, vitality and decline, self-identities and social representations of the aged. It is increasingly recognised in gerontology that the 'kaleidoscopic' (Hazan, 1994: 3) character of the realities, experiences, times and places of old age defies conventional and uni-disciplinary attempts to understand it (Katz, 1996). It is now argued that the myriad personal, existential, cultural and political issues involved in ageing, as well as the plural, contradictory and enigmatic meanings of ageing, require the pioneering efforts of those who cross disciplinary boundaries and engage in multiple areas of knowledge production. It is a call for what Fabian has referred to as 'border crossing . . . between science and literature, ethnography and biography, interpretation and imagination' (1993: 52).

This chapter begins by tracing the emergence of gerontology as a unified science in order to contextualise the efforts of scholars who, through the development of multi-disciplinary configurations in gerontology, have sought to expand our understanding of ageing beyond a unified discipline. The impetus for these developments is to escape the reductionism of a unified science. As Katz states, gerontologists who are increasingly critical of social gerontology for its 'narrow scientificity, advocate stronger ties to the humanities, endorse reflexive methodologies, historicise ideological attributes of old age, promote radical political engagement, and resignify the ageing process as heterogeneous and indeterminate' (1996: 4). Emerging from the work of such critics is what Achenbaum has termed 'inter-disciplinary amalgams' (1995) through which scholars break from the positivist tradition of gerontology to make visible the complexity and contingency of old age while resisting the temptation to produce uniformity (Gubrium, 1993b). After tracing the emergence of gerontology as a unified science I explore three such amalgams – humanistic gerontology, anthropological gerontology, and critical cultural gerontology – focusing on how ageing in place and the temporal conditions of growing older are conceptualised in these inter-disciplinary and boundary-crossing developments.

Gerontology as a unified science

Gerontology's founders recognised early on that ageing was interactive and multi-faceted, which confirmed the importance of taking a multi-disciplinary approach to understand the problems of ageing. However, despite the multi-faceted nature of ageing, it was assumed that all its aspects were interrelated such that gerontology could evolve into a unified field of enquiry with a 'homogeneous medium' (Achenbaum, 1995: 75) for the articulation of diverse perspectives on ageing. As Achenbaum writes (1995: 78), the founders of gerontology sought to bridge the natural and social sciences and 'envisioned that the basis for consensus in research on ageing would rest on high standards of scientific excellence'. Approaches generated by the micro-perspectives of biology, chemistry, immunology and other 'hard sciences' were embraced in order to achieve this consensus.

With the impetus in gerontological investigations towards searching for scientific laws and adhering to scientific methodology, the discipline's technical or instrumental reasoning ignored what could not be reduced to quantitative methodology. Yet with vast research areas totally unexplored it became increasingly apparent that the quest of gerontology to become a unified science created a hierarchy whereby biomedical research became the pervasive force in the shaping of gerontological multi-disciplinarity (Achenbaum, 1995).

The vision of gerontology as a unified science became difficult to sustain with the vast and rapidly growing accumulation of empirical data that could not be assimilated to any single unified theory or bounded and managed by a single discipline (Cole, 1993). Further, there was the increasing acknowledgement among gerontologists that the exclusive focus of geriatricians on physiological changes produced a limited social construction of ageing. Because the very complexity of ageing created what Klein has referred to in her analysis of the literature on inter-disciplinarity as 'a sense of inter-disciplinary necessity' (Klein, 1996: 40), exploring perspectives from multiple disciplines and engaging in multiple boundaries of knowing was proposed as a resolution to these epistemological problems with which contemporary gerontology was beset.

Humanistic gerontology

The concern to reinsert human subjectivity into the study of ageing has prompted an interest in qualitative and interpretive methods in ageing research. The past decade has marked an upsurge of interest in humanistic approaches to the study of ageing where, for example, literature is used as a source of ideas about the meaning and experience of ageing (Hepworth, 2000). Literary gerontology takes as its focus the interpretation of ageing and creativity through close readings of literary texts (Basting, 2000). In highlighting the tension between scientific findings and 'real-life dramas' (Kastenbaum, 1994), it is argued that literary representations serve as a vital source of ideas about cultural representations of ageing and their influence on individual subjective experiences including expectations that individuals hold about the ageing process (Hepworth, 2000). Fiction is treated as a valuable resource because 'it allows the writer, through the exercise of imagination, access to the personal variations and ambiguities underlying the common condition of growing older' (Hepworth, 2000: 4). Fiction is not imaginary, but imaginative (Woodward, 1991), and thus the exercise of creating fictional moments allows the author to depict human nature richly, evoking, resembling and illuminating human reality with insight and depth. Ignatieff, in discussing why he wrote about his mother's struggle with Alzheimer's as fiction, says that working his way into the fictional thread of his novel *Scar Tissue* allowed him to develop scenes born from his imagination. Reflecting on this process in an interview, Ignatieff notes that 'an enormous discovery for me . . . [was] that something that I had imagined could be truer or could say more about what I felt at the time than what had actually occurred' (2002). Thus Ignatieff was able to delve into the mysteries of Alzheimer's disease and contemplate difficult questions about science, genetics, art and personhood, limited only by the boundaries of his own imagination.

Fiction has explored the complicity between identity and the places and spaces where the elderly live. Places serve as crucial material and symbolic sources for biographical development and, as such, make an essential contribution to the construction of personal identity. Attachment over time to, for example, homes and their community surroundings serves as a crucial experiential anchor for memories, personal histories and narratives, and it is in this respect that we can say that 'emplacement' is linked to personal identity (Hepworth, 2000: 78). For example, in Hepworth's exploration of fictional descriptions of ageing, he refers to Cobbold's novel *A Rival Creation*, which illustrates the reflexive interaction between place and identity. In the novel, Evelyne Brooke, an elderly nature conservationist who diligently tends to and takes pride in her flourishing garden, is described as having aged overnight when her garden is vandalised by a vengeful neighbour. As Hepworth notes, Evelyne's sudden and rapid deterioration is indicative that the garden was 'the complete external expression of her inner self' (2000: 77). In other words, the destruction of the garden was a symbolic assault on her personal identity manifested in physical decline associated with ageing.

One of the strongest formative influences of place on identity, and perhaps the most emotive, is the home (Kontos, 1998; Hepworth, 2000). Relph, one of the first geographers to examine sense of place, devotes a portion of his book to the notion that home places are 'profound centres of human existence' (1976: 39) in that they provide 'not only the context for all human activity, but also security and identity for individuals and groups' (1976: 41). This is apparent when we consider the move from a private home to a nursing facility, a transition that is often described in gerontological literature as one of loss and regret (Groger, 1995). Faced with the prospect of losing one's home is an extremely painful experience for both elderly individuals and their families and, as Hepworth notes (2000: 94), home is thus 'a prominent narrative turning point in stories of ageing'. For example, Ignatieff's *Scar Tissue* describes the anger that the female protagonist who struggles with advancing Alzheimer's expresses when her son admits her into a nursing facility. As the son begins to unpack from her suitcase the few material possessions her small room is able to accommodate, she stares out of the window, absently smoothing the coverlet of the bed with her hand. The last thing that he does is carefully remove her wedding band, a safety regulation enforced by the nursing home. He then stands up from the bed and says 'goodbye', assuring his mother he will be back later. Suddenly the mother emerges from her seeming absence, crying out 'No, no, no', and then, with tremendous emotion and conviction, she yells, 'I can't stand it. I can't' (Ignatieff, 1993: 99). The reader is struck by the awareness that not even the loss of cognition could ease the pain and fear of losing one's home and familiar surroundings.

An important aspect of fictional representations of the complicity between identity and the places where older individuals live is the symbolic role of 'biographical objects' in shaping identity (Hoskins, 1998). Following Hoskins, identities and biographies are formed around objects such as photographs, jewellery and other personal memorabilia; so significant are biographical objects in the events of an individual's life that they can become 'surrogate selves' (1998: 7). The notion that objects become a vehicle for a sense of self is echoed by Woodward, who states that 'the elderly invest themselves in souvenirs of their pasts which serve as emblems of their present selves' (1991: 137). It is in this context that Woodward explores Beckett's *Malone Dies*, a fictional account of an old dying man who remains connected to the world through physical contact with a small inventory of his own possessions. Confined to a bed in a somewhat barren room in a large house, the infirm and immobile Malone clings to specific objects which speak to him about his past. As Woodward comments (1991: 136), 'his objects testify to his present existence . . . and his periodic rummaging in them offers the consolation of the sense of continuity of the self'.

It is important to note that humanistic gerontology involves more than the textual focus of literary gerontology. There is as well a growing literature in arts-informed approaches and on the use of various representational forms in gerontological research (Kastenbaum, 1994; Basting, 1998; Cole *et al.*, 2000; Ray, 2000). Performance studies of age have their roots in multiple disciplines, such as theatre studies, folklore, anthropology and sociology. Among the rare early depictions of ageing and performance is the work of anthropologist Myerhoff, who studied the members of the Aliyah Senior Citizens' Center in Venice, California, a group of elderly migrants from Eastern Europe. She explained their use of performance through ritual gestures to affirm their identity. Myerhoff stresses the significance of the members' enactment of ancient long-known ethnic ritual gestures such as reciting the candle blessing on the sabbath, one of the sacred Jewish holidays. She argues that, originating in the most basic layers of childhood, the earliest emotions and associations of domestic life, these rituals 'arouse deep involuntary memories, those surges of remembrance that are not merely accurate and extensive but bring with them the essences and textures of their original context, transcending time and change' (1978: 225). Thus we see in Myerhoff's analysis of these 'definitional ceremonies' (1992) the significance of place, namely the home, which is redolent of early, fundamental associations. It is Myerhoff's assertion that rituals associated with the earliest social experiences have potent emotion and cultural rootedness and, as such, have the capacity to carry one back to earliest times with nostalgia and great yearning (1978). This is aptly captured by Basha, one of the members of the Aliyah Center, who explains to Myerhoff what it meant to her to say the prayer over the Shabbat candles:

> When I was a little girl, I would stand this way, beside my mother when she would light the candles for Shabbat. We were alone in the house, everything warm and clean and quiet with all the good smells of the cooking food coming in around us . . . To this day, when the heat of the candles is on my face, I circle the flame and cover my eyes, and then I feel again my mother's hands on my smooth cheeks.
>
> (Myerhoff, 1978: 256)

Myerhoff discusses how the sabbath is powerfully associated with family and household and it is this blending of nurturance and ethnic specificities that explains why the sabbath ritual affected Basha and stayed with her. Reciting the candle blessing aroused a kind of 'experiential remembering' (1978: 226) of household smells, sounds, tastes and sentiments, a mixture that is an accumulation of historical moments in time and place. 'Experiential remembering' of the meaning or experiences of home that is elicited by the performance of ritual is also evident if we consider the statement made by Moshe, another member of the Aliyah Senior Citizens' Center, who describes his experience reciting the Kaddish (the prayer for the dead):

> Whenever I say Kaddish, I chant and sway, and it comes back to me like always. I remember how it was when my father, may he rest in peace, would wrap me around in his big prayer shawl. All the warmth comes back to me like I was back in the shawl where nothing, nothing bad could happen.
>
> (Myerhoff, 2000: 436)

As Myerhoff elaborates in an earlier publication (1978: 225), Moshe experienced unification with his father as he recited the Kaddish, reliving the first time he had heard the prayer and once more feeling himself as a small boy wrapped snugly in his father's prayer shawl while weeping and swaying over an open grave. Physical security, relational affection and emotional security are aspects of the experience of home that clearly resonate as Moshe recites the traditional Hebrew prayer.

Anthropological gerontology

An important area where anthropologists have contributed to the theoretical development of gerontology is ethnographic descriptions[1] of the internal relationships and subjective conditions within institutional, residential and community-care spaces. Ethnographies of 'spaces of age' (Katz, forthcoming) include private homes (Rubinstein, 1986; 1990b), supportive housing buildings (Kontos, 1998), retirement communities (Katz, forthcoming), nursing homes (Gubrium, 1975; 1993a; Hazan, 1994; Henderson and Vesperi, 1995), community centres (Myerhoff, 1978), Universities of the Third Age (U3A) (Hazan, 1996), and support groups (Gubrium, 1986). Ethnographers of ageing seek to understand the places associated with the elderly by experiencing those places from within, participating in local practices while simultaneously observing and analysing those practices. As such, it has the advantage of drawing the observer into the phenomenological complexity of a particular social and cultural world,

> that domain of everyday . . . with all its habituality, its crises, its vernacular and idiomatic character, its biographical particularities, its decisive events and indecisive strategies, which theoretical knowledge addresses but does not determine, from which conceptual understanding arises but on which it does not primarily depend.
>
> (Jackson, 1996: 7–8)

Diamond (1992) and Henderson (1994) are ethnographers who studied the culture of the work of nursing assistants in nursing homes. Going beyond the usual method of participant observation in their respective studies, they themselves trained and worked as nursing assistants in a nursing home in order to fully understand this work. Their research is based on the premise that in order to improve conditions at nursing homes one must understand not only the illness of the residents there, but the stresses and strains of care-providers.

While the ethnography of nursing homes is a growing field of study (Henderson and Vesperi, 1995), with a few noted exceptions (Chatterji, 1998; Vittoria, 1998; Reed-Danahay, 2001) researchers have rarely focused on Alzheimer's care. Reed-Danahay's ethnographic work in a dementia unit is noteworthy. She explores the dislocation of Alzheimer's patients highlighting how the social and spatial arrangements of the dementia unit prompted many residents to confuse contexts of work and home, thus believing that they were at work instead of at the nursing home. She suggests that rather than viewing the patients' dislocation as a symptom of dementia, this confusion, or perhaps 'nostalgia for home and work' (Reed-Danahay, 2001: 60), should be understood as being, at least in part, a response to the bureaucratic setting in which they were living. The key to understanding the patient's dislocation, according to Reed-Danahay, is recognising that despite our contemporary yearning for nursing homes to be 'home-like', dementia units are not 'homes' but rather are 'non-places', characterised by relative anonymity where Alzheimer residents are strangers interacting with each other, 'placed together through circumstance rather than volition' (2001: 50).

In increasing numbers social scientists are questioning the reliance upon the medical model of Alzheimer's which overlooks the impact of treatment contexts and care-giving relationships on disease progression. The social and interpersonal nature of the illness is central to this research as are concepts such as 'the local culture of dementia' (Kitwood, 1993), the social construction of senility (Gubrium, 1986; 1987), and processes such as 'infantilization' (Lyman, 1988: 72), 'labelling', 'stigmatization' and, ultimately, 'banishment' (Kitwood, 1990: 182–3). Focus on micro processes of interaction in the context of everyday life has brought the social forces involved in dementia to the fore, challenging the construction of the disease as a fixed biomedical reality. Experiences of the illness are situated in a biographical context which has served to bring into sharp focus those interactions that depersonalise, disempower and invalidate a sufferer of dementia. This research illuminates the impact of cultural definitions, care settings and care-giving relationships on the experience of dementing illness as well as disease progression. The culmination of these developments is a move towards recasting the disease as primarily the result of a sociological process rather than of a strictly neurological and behavioural process.

Although not part of the ethnographic enterprise, the work of Kitwood (1990; 1993; 1997) on the impact of non-biomedical factors on the experience and progression of dementia is worthy of mention here given the prominence that place holds in his assertions regarding dementia. Rather than viewing dementia solely as a manifestation of brain pathology, Kitwood has proposed that cultural, organisational and interactional dynamics in particular settings can provoke reactions, both positive and negative, in the sufferer (Kitwood, 1998). Central to Kitwood's position is that labelling and treating individuals as demented leads to a type of 'malignant social psychology' (Kitwood, 1990: 181) that damages their fragile self-esteem and personhood, leading ultimately to the loss of self that is so widely thought to be caused by neuropathology. Traditional approaches to care-giving are largely informed by the assumption that with cognitive impairment there is a corresponding erosion of the self, a process that has been referred to as 'unbecoming a self' (Fontana and Smith, 1989: 36) which leaves only a body from which the person has been removed (Keane, 1994: 152). This assumption is perhaps most apparent when we consider the depersonalising tendencies of some treatment contexts and care-giving relationships in which individuals are treated as demented (Kitwood, 1990; 1997). As Marshall notes in her discussion of care environments, it is not unusual to see staff treat residents with dementia as 'neither needing nor deserving even the normal rules of decent human interaction' (2001: 175). Such care environments undermine personhood and, as such, are regarded as constituting a 'malignant social psychology', literally a 'cancer of the interpersonal environment' (Kitwood, 1998: 29). Kitwood is careful to avoid the suggestion that there are any malicious or deliberate intentions on the part of care-givers to harm dementia sufferers. On the contrary, he notes that interactions that have a negative impact on the person with dementia are a consequence of ignorance and preoccupation, but above all of bad practice that is part of a deeply embedded cultural tradition (Kitwood, 1998).

Kitwood's work has fostered an awareness of the cultural content of Western assumptions about dementia as well as a critique of Western forms of treatment and response. The cultural construction of Alzheimer's is perhaps most powerfully illuminated in cross-cultural analyses that localise the normative content of Alzheimer's discourse. Cohen's ethnographic research in Banaras, India, explores cognitive and affective change 'in terms other than or in addition to those of a diagnosis or a disease: as a set of *local* and *contingent* practices rooted in culture and political economy' (Cohen, 1998: 6, emphasis in original). Cohen describes senility, what is most commonly spoken of in Banaras in terms of 'weak brain' (1995: 318), as a physical consequence of ageing but emphasises that for the Banarsis, this is 'inadequate as a basis for understanding what is at issue in such debility' (Cohen, 1995: 318). It goes beyond the scope of this chapter to fully elaborate the explanatory model of 'weak brain'; however, what is of significance for my discussion of ageing and place is that the meaning of 'weak brain' cannot be understood divorced from the politics of domestic space and inter-generational life. More specifically, 'weak brain' is in part understood as a consequence of children not offering their parents adequate support or respect. Inter-generational life and exhortations to honour one's parents are so intimately bound to domestic space that parents are closely identified with the household and with authority over it. As a consequence, according to Cohen, 'weak brain' threatens the moral integrity of the family, and in this respect points to the limits of a neurological model of senility in Banaras.

Ethnographic research on the local cultures of 'spaces of age' (Katz, forthcoming) brings the human subjectivity of ageing to centre stage and explores how place has significant implications for how old age is experienced. Further, ethnographic studies provide a powerful corrective to the tendency in mainstream gerontology to understand older people as empty vessels merely reacting to, rather than shaping, experiences and spaces of ageing in later life. More and more we are seeing how, for example, supportive housing buildings (Kontos, 1998) and day centres (Hazan, 1983; Myerhoff, 1978) are being shaped and utilised in innovative ways by the elderly themselves rather than administrators and staff. For example, in an ethnographic study of Home Frontier, a Canadian supportive housing building, I explore the complexity of home as a place that 'facilitates independent daily living, and as a local construction that is negotiated and contested in the practice of its use' (Kontos, 1998: 168). Rules and regulations which are enforced by the staff of the building are perceived by the tenants as features of institutional living which compromise the independence of the tenants and effectively threaten their personal identities (1998: 182). Home Frontier is thus a struggle between a vision of 'home-as-place and place-as-institution' (1998: 168), a struggle that is dramatised by the tenants' persistent efforts to negotiate with the administrators of the building to make it into, and maintain it as, the kind of community they want. Hanging framed prints of some kind in the lobby, mounting a mirror, redoing the wallpaper in the tea room, and installing brighter lights were among the most common suggestions made by the tenants to make Home Frontier look less institutional and more like home. Tenants also organised social programmes that better satisfied their interests as opposed to those implemented by the programming staff. Euchre, bingo, pub night, singing classes and the Breakfast Club – an open drop-in at which coffee, tea, toast and muffins were available for all tenants for the nominal cost of one dollar – were among the programmes that were organised and run by the tenants themselves. The clash between staff's conception and treatment of Home Frontier as an administrative territory, and the tenants' insistence that it is a home community, is powerfully illustrative of how home, as a physical place and a meaningful context for everyday life, not only facilitates the tenants' independence but, in so doing, also sustains their personal identity.

Ethnographic research is also exposing the limitations of traditional spaces, such as educational environments of the elderly, which are themselves being challenged and reordered as in the case of the Elderhostel movement and Universities of the Third Age (U3A) (Katz, 1996). Katz, in his discussion of these organisations, notes that they have collectively restructured the experience of receiving education: Elderhostel facilitates visits to universities, national parks and historical sites to provide educational experiences that combine courses, retreats and excursions and U3As have transcended the division between instructor and student in that everybody is an equal member in a collaborative process of learning, there are no requirements for formal qualifications and exams, and classes are often held in members' homes. As Katz asserts, Elderhostel and the U3As are indicative of the multi-spatial dimensions of ageing and of how social space is becoming reconfigured in function, opportunity and meaning by the 'emergent energies of the "Third Age" itself' (1996: 138).

Critical cultural gerontology

Critical cultural gerontology is the convergence between critical gerontology and cultural studies, an inter-disciplinary approach that is advocated as a way of strengthening ageing studies and moving ageing into a more prominent position conceptually (Gullette, 2000). Following Gullette's explication, both these approaches are 'influenced by feminist, poststructuralist, multicultural, and left theories . . . [and] both share a commitment to examining cultural practices, economic conditions, and public policy from the point of view of their involvements with power' (2000: 217). At this intersection, there is also the shared commitment to critical gerontology's interest in examining the linkages between discourses, social processes and power that are generally opaque, and fostering critical awareness of language and discourse that is necessary to render them transparent. The focus on the taken-for-granted and relations of domination opens the space for seeing that things could be otherwise and for envisioning new possibilities (Moody, 1993).

In the spirit of an 'information society', a 'postindustrial society', and a 'postmodern culture' (Moody, 1993), in an emergent consumer culture discourse 'elderhood has been reconstructed as a marketable lifestyle that connects the commodified values of youth with bodycare techniques for masking the appearance of age' (Katz, 1995: 70). Instrumental strategies for body maintenance, such as pharmaceutical drugs, cosmetic surgery, dieting and exercise, resonate with the strength of consumer culture to create the desire for health, longevity, sexual fulfilment, youth and beauty. Within consumer culture the older body is rejected as a possible site of beauty. As Sceats argues, 'beauty and age are seen as incompatible, the beauty *of* age an oxymoron' (2003, emphasis in original). However, the malfunctioning body can be repaired, holding at bay the ravages of time and 'turning old age into an extended active phase of "mid-lifestyle"' (Featherstone, 1995: 228). Marketing strategies aimed at 'ageless' consumers (Featherstone, 1995; Featherstone and Hepworth, 1991) provide an array of accoutrements, ranging from various regimes and techniques of body maintenance to increasingly sophisticated medical technology, promising the prospect of erasing boundaries between chronobiology and physiological capacities in old age. Katz and Marshall (2003), for example, have engaged in a critical exposure of how sexuality has emerged as a pivotal concern for rehabilitating the ageing body. Resting on the recent 'cultural-scientific conviction that lifelong sexual function is a primary component of achieving successful aging' (2003: 12), Katz and Marshall discuss how refashioning sexual dysfunctions as age-related organic diseases, rather than as a natural consequence of ageing, is part of the postmodern ideal that people age outside of time. Liberating sexual performance from the limits of the ageing body, it is argued, creates the medical imperative for consumers to take responsibility to ensure their continued sexual capacities. This renders any unwillingness to accept this medical imperative irrational. Thus, as Katz and Marshall argue, 'individuals skilled in the exercise of consumer choice must choose to be posthumanly and timelessly functional and do so by embracing the benevolent promise of modern technology' (2003: 13).

The proliferation of images within consumer culture has not only restructured the life course through a destabilising of chronological boundaries but has also affected the material fabric of everyday life through a restructuring of social space. Built environments for 'Third Agers' (Blaikie, 1999: 175) are increasingly reflective of the new social ageing and its idealistic imagery. Thus the importance of consumer culture for my discussion of ageing and place is that with the shift to a postmodern life course there has been a corresponding cultivation of what Blaikie refers to as 'landscapes of later life' (1999). For example, as Blaikie comments, the 'land of old age' is being redeveloped such that traditional old age homes are becoming marginalised relics in the face of the development of specialised housing such as congregate care complexes which often emulate the detail and style of grand-deco hotels and also offer a 'hotel lifestyle' (1999: 176). New retirement developments such as Sun City retirement communities are also part of the new landscape of old age. These communities, as Katz argues (forthcoming), often publicised as 'escapes' and 'havens', have little to do with how ageing is more commonly experienced and thus they effectively mask the ageing process by naturalising retirement living as continuously active, healthy and problem-free.

In this culture of timelessness, technological reconstruction of the body is not the only way in which chronological age is being discredited as the 'true index of how a person should feel' (Featherstone and Hepworth, 1991: 374). Even for those individuals who live with severe bodily decline through disabling illness, who often experience the outer biological degeneration of the body as imprisoning their inner youthfulness (Featherstone and Hepworth, 1991; Turner, 1995), developments in communication media have created the possibility to tentatively escape from the here-and-now embodiment of the everyday world. More specifically, computer-generated virtualities provide the opportunity to escape from the 'body as prison' and experience a 'freedom and mobility in sensorily exciting worlds, such as simulated mountain climbing' (Featherstone, 1995: 237). As Featherstone describes, 'telepresence' (1995: 235) facilitates the experience of shared 'nonplace' worlds whereby simulated environments can be created in order to provide spaces for disembodied interactions and the ability to express a range of experiences (1995: 233). Thus virtual reality offers highly realistic out-of-body experiences that create, even in the case of severe disability, the impression that one is in another place moving and interacting with the full range of human senses.

Conclusion

Multi-disciplinarity is a fundamental intellectual resource by which gerontology has grown as a profession since the early twentieth century (Achenbaum, 1995). Vibrant sub-fields that blend diverse approaches, plural knowledges and shared expertise have created a hybrid literature incorporating ideas from the social sciences, humanities and cultural studies. 'Cross-fertilizations of borrowing' (Klein, 1996: 61) interpretive methodologies and cross-cultural frameworks have been a vital stimulus for challenging the instrumentalism of mainstream gerontology, rendering taken-for-granted categories and claims to universality vulnerable to displacement and deconstruction. Through the development of inter-disciplinary amalgams such as those explored in this chapter, gerontologists have engaged in progressive knowledge-building by broadening ageing studies beyond the narrow reductionism of biomedical frameworks to encompass the heterogeneity of late life and the indeterminacy of the boundaries and experiences of old age.

However, as Katz alerts us, multi-disciplinarity in and of itself should not be seen as a panacea since 'hybrid developments . . . are themselves in danger of becoming stagnant and disciplined without continual rehybridizing' (1996:

110). In other words, as Katz argues, specifying and enriching the critical strand of gerontological knowledge depends less on the organising and ordering practices of multi-disciplinarity, and more on the innovative possibilities created by 'intellectual fragmentation' (1996: 141; see also Katz, 2003). The suggestion here is for gerontology to articulate its criticality via the 'continual recombining of plural, dislocated, interdisciplinary fragments' (Katz, 1996: 140). With the new construction of the life course and the redevelopment of spaces of ageing, together with the increasingly diverse and unpredictable nature of ageing, it seems that the intellectual enterprise of gerontology can only gain by approaching the study of ageing through the prism of inter-disciplinary fragments.

Note

1 Though ethnography was originally the province of anthropology, it is a method that is employed by sociologists and other social scientists as well. In being utilised by sociologists, this boundary-crossing in the realm of methodology has enriched ethnographic explorations of ageing.

4

INTERNATIONAL DEMOGRAPHIC TRANSITIONS

Kevin McCracken and David R. Phillips

Introduction: demographic ageing as a worldwide phenomenon

Population ageing, sometimes called demographic ageing, is now a worldwide phenomenon. Technically any population whose average age is rising can be said to be ageing, but usually the term is more specifically used with reference to an increase in the proportion of persons aged over 60 or 65.[1] Such an increase has been taking place in the currently developed regions for well over a century and is now also increasingly evident in the countries of the less-developed regions. Today we tend to think of the demographically older countries as comprising principally those in Europe and a few countries elsewhere. However, already well over half (59%) of the world's older people are found in developing nations and by about 2050, of the expected 1.4 billion persons aged 65 and over worldwide, almost 80% (1.1 billion) are likely to be in the currently less developed regions. Barely 20% of the world's older persons will by then live in what we today call the developed regions (Table 4.1).

Demographic ageing is clearly a worldwide phenomenon, noted by many agencies and summarised by the United Nations (2002a), in preparation for the Second World Assembly on Ageing held in Madrid that year. The next 50 years or so are likely to see the following key demographic facets:

- The global population of older persons is growing by about 2% annually, much faster than the population as a whole and this rate is expected to increase to as much as 2.8% per annum by 2025–30.
- There will be very marked differences between the world's regions in numbers and proportions of older persons. In the more developed regions, the proportion of people aged 65 and over will increase from 14.3% to around 26% between 2000 and 2050. In the less developed regions, this growth will be from about 5% in 2000 to about 14.3% in 2050, or the same percentage as are aged 65 and over in today's more developed regions.

Table 4.1 Population aged 65 and over, and 80 and over, world and regions, 2000 and 2050

	Population aged 65 and over				*Population aged 80 and over*			
	Millions		*% of total population*		*Millions*		*% of total population*	
	2000	*2050*[1]	*2000*	*2050*[1]	*2000*	*2050*[1]	*2000*	*2050*[1]
World	419	1419	6.9	15.9	69	377	1.1	4.2
More developed regions	171	316	14.3	25.9	37	113	3.1	9.2
Less developed regions	248	1103	5.1	14.3	32	265	0.7	3.4
Africa	26	122	3.2	6.8	3	20	0.4	1.1
Asia	216	880	5.9	16.8	29	224	0.8	4.3
Europe	107	177	14.7	27.9	21	60	2.9	9.5
Latin America and Caribbean	28	140	5.5	18.2	5	38	0.9	4.9
Northern America	39	92	12.3	20.5	10	33	3.2	7.4
Oceania	3	9	9.8	19.1	1	3	2.2	6.1

Source: United Nations, Population Division (UNPD) (2003) *World Population Prospects: The 2002 revision population database* (http://esa.un.org/unpp).

Notes
1 Medium variant projection.
Regional totals may not add up to world total due to rounding.

- The rapid pace of population ageing will make it difficult for developing regions in particular to adjust their economic and social care systems.
- The older population is itself ageing and the oldest-old group, aged 80 and over, is currently increasing at about 3.8% per annum and comprises about one-sixth of older persons, but this will reach more than one-quarter by mid-century. Indeed, the percentage of total population aged 80 and over will probably increase from 1% to 4% by 2050.
- Population support ratios (regardless of their limitations) indicate that there will be a growing dependency burden of older persons on younger workers. These effects will be important for social support and social security, but will be very much affected by the wealth and health status of the older generation.

There are many other important demographic changes identified by the United Nations and numerous other official and academic reports (see for example WHO, 2002; Kalache and Keller, 2000; Kinsella and Phillips, 2005). These publications point strongly to the importance of identifying both geographical and socio-economic variations in patterns of ageing. An especially important issue for future policy towards older persons will be their health status. This is very much associated with cohort trends in health, generally improving with successive age groups, and the global patterns of epidemiological transition discussed below.

Demographic and epidemiological transitions and population ageing

Why do populations age?

The major underlying cause of demographic ageing is falling fertility (Coale, 1964; Leete and Alam, 1999; Kinsella and Phillips, 2005), although the public perception is that the principal cause is increasing life expectancy. The latter does certainly play a part but, in fact, the most influential factor in population ageing has been changing fertility, declining birth rates causing the proportion of the population at younger ages to contract and, concomitantly, the proportion at older ages to expand (i.e. 'ageing from the bottom'). In this way, the explanation of population ageing in late-stage demographic transition has hinged on people's long-term reproductive behaviour – crucially, the decision to have far fewer children.

Rising life expectancy through combinations of social and medical advances has led to greater numbers of people surviving into their 60s, 70s, 80s and higher, but to date has had little impact on the proportional strength of the elderly in most populations as mortality improvements have also benefited those at younger ages. The scope for further mortality decline from infancy up to late middle age, however, is becoming limited in the more developed nations. For example, survivorship to age 65 is now around 89% in Japan, 88% in Australia, 86% in the United Kingdom and 82% in the United States. Hence virtually all future improvements in mortality will be at older ages and will become the major contributor to population ageing (i.e. 'ageing from the top'). In recent years the shift from fertility-dominated to mortality-dominated ageing has in fact occurred in a number of very low-fertility nations (e.g. France, Italy, Japan).

Total fertility rates (TFRs, the average number of children that would be born per woman of reproductive age if she conformed to the age-specific fertility rates of the period) have certainly steadily dropped almost everywhere, but at a variable pace. A century ago in today's developed countries five or six children per woman were commonplace, whereas today two or fewer have become the norm. Family sizes are likewise dropping across the developing world, though some 'pre-transitional laggards' remain, 17 countries (15 in Africa, plus Afghanistan and Yemen) being identified in the most recent United Nations revision of world population prospects as still having TFRs of six or seven (UNPD, 2003) (see Figures 4.1 and 4.2).

The largest declines in fertility within the developing world have been in Asia, where numerous countries that only a few decades ago had TFRs of six or more children now have rates below or approaching replacement level – e.g., Singapore (1.4), South Korea (1.4), China (1.8), Thailand (1.9), Sri Lanka (2.0), North Korea (2.0), Vietnam (2.3), Iran (2.3) and Indonesia (2.4). In turn, many former high-fertility countries in Latin America have TFRs under three, some substantially below, such as Brazil (2.2), Costa Rica (2.3), Mexico (2.5), Chile (2.4) and Colombia (2.6), all having fallen drastically since the 1970s and 1980s.

By the United Nations' calculations (UNPD, 2003), in 2000–5, some 62 countries had TFRs below the replacement rate of 2.1 children, at which parents reproduce themselves, so technically their populations will eventually shrink as more people will die than are born. A substantial number of these countries, among them Hong Kong, Bulgaria, Russian Federation, Spain, Hungary, Italy, Poland, Greece, Austria and Japan, in fact have TFRs as low as 1.0 to 1.3. Such persistent low fertility over the past couple of decades has led to a decline in the size of successive birth cohorts and a corresponding increase in the proportion of the older relative to the younger population. Government attempts in some countries to raise the birth rate by social and financial policies have only had a minimal impact, so it seems that low fertility rates, once established, are difficult to turn around.

Indeed, as Kinsella and Phillips (2005) note, a recent study opens with a statement that many would have found unbelievable just a decade earlier: that a majority of the global population live in countries with near- or below-replacement fertility (Kohler *et al.*, 2002). That a country with a huge population such as China now has a TFR of about 1.8 contributes to this phenomenon. However, as fertility change in most low- and middle-income countries has been relatively recent and rapid, basically only occurring since about the 1980s, the aggregate populations in such countries will continue rising for some years as many people are now in the childbearing and pre-childbearing years. That is, momentum for further substantial growth is built into their current age structures. To take Thailand as an example, replacement-level fertility was reached in the early 1990s, at which time the national population was 56 million. If the country follows the United Nations' medium-variant fertility-decline scenario, population growth will continue until the mid-2040s, reaching around 77 million. More rapid fertility decline would see the population peak earlier, but still at many millions more than the current total.

For population ageing, in general, populations with high fertility tend to have low proportions of older people and vice versa. The well-known term *demographic transition* refers to the gradual process by which societies move from high fertility and mortality rates to lower rates of both (see Table 4.2 and Figure 4.5). The initial phase of the process is usually a decline in infant and childhood mortality, as infectious and parasitic diseases are controlled. Fertility rates tend to remain high for one or two generations before the improved infant and childhood survivorship makes a significant impact on couples' reproductive decisions. Birth rates in fact sometimes even increase in the early part of this phase due to health improvements in females. This conjunction of lower infant and childhood mortality and continuing high fertility produces large birth cohorts and an expanding proportion of children relative to adults, resulting in an initially even younger population age structure (Lee, 1994). However, longer-term, as birth rates fall and health levels further improve through social, medical and public health advances, a demographically ageing population emerges.

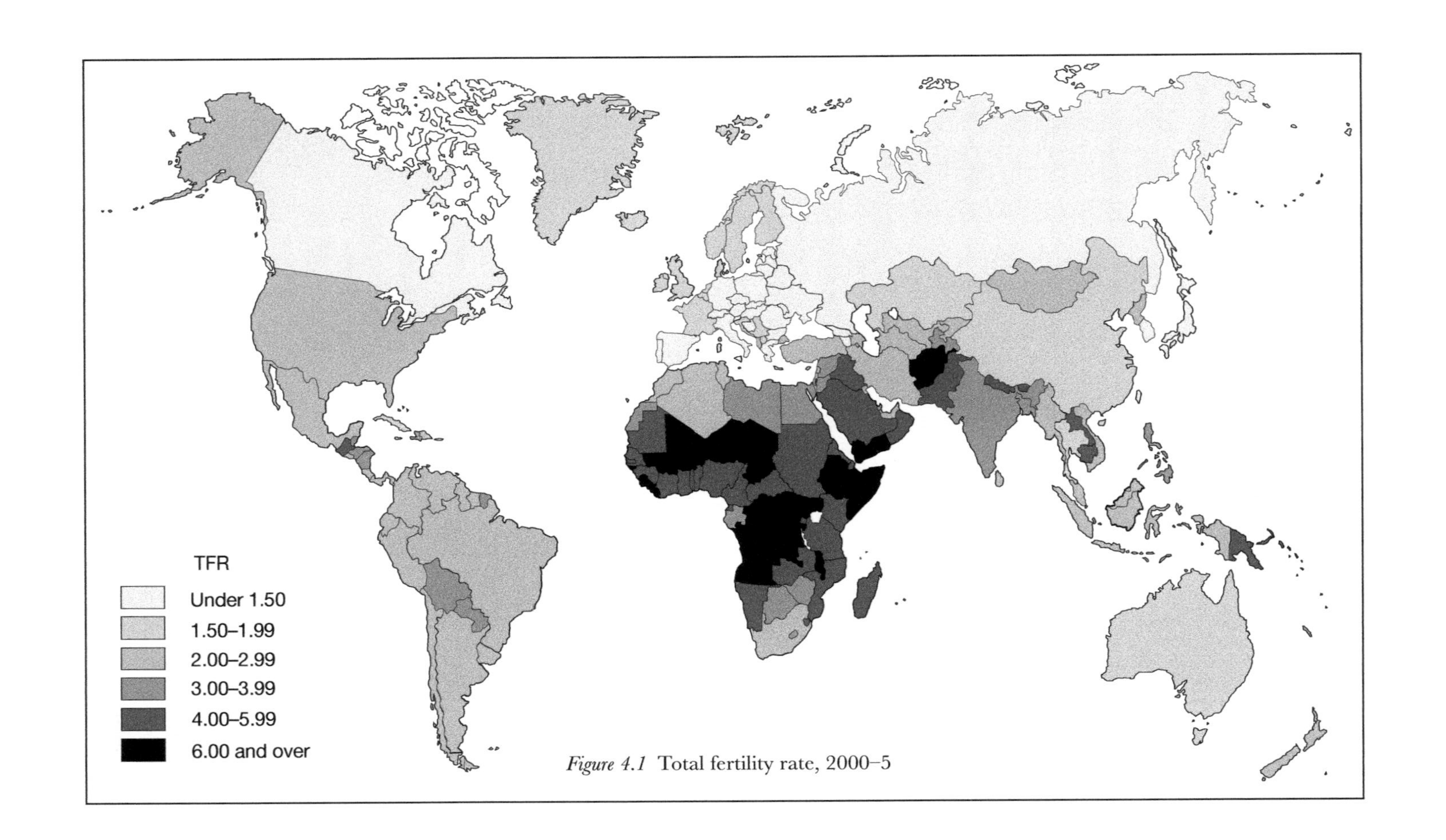

Figure 4.1 Total fertility rate, 2000–5

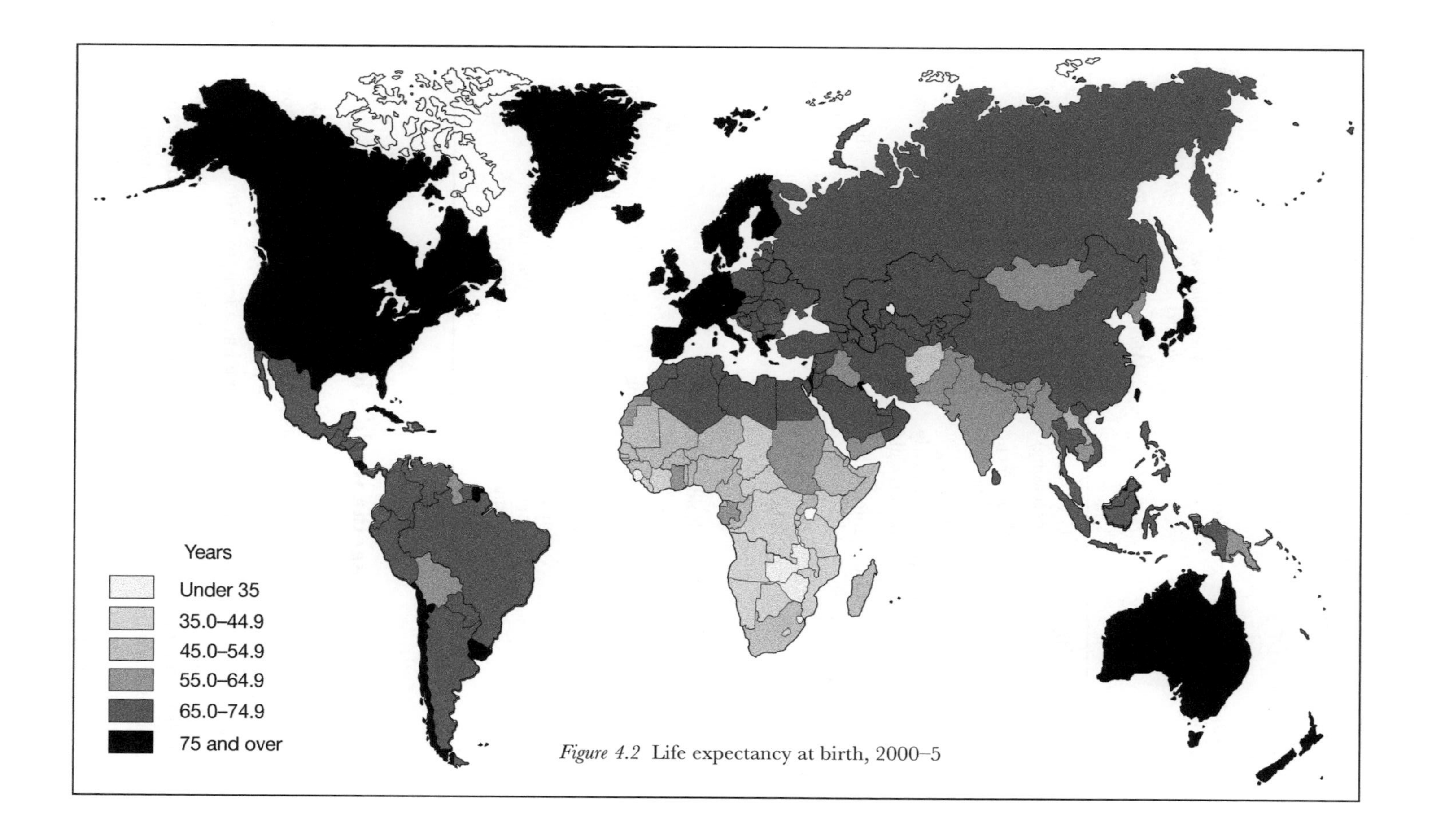

Figure 4.2 Life expectancy at birth, 2000–5

Table 4.2 Summary of demographic, epidemiological and ageing transitions

Stage of Demographic transition	*Pre-*	*Early*	*Late*	*Post-*
Epidemiological transition	*Pestilence and famine*	*Receding pandemics*	*Degenerative and human-induced diseases*	*Delayed degenerative diseases and emerging infections*
Ageing transition	*Young*	*Youthful*	*Mature → Aged*	*Very aged → Hyperaged*
Fertility	High	High (perhaps increasing) Subsequent initial decline	Declining	Low (sub-replacement?)
Mortality	High (and fluctuating)	Declining (particularly infant/child)	Continuing decline	Low (improvements at older ages)
Main causes of death	Infectious diseases	Infectious diseases declining Degenerative diseases (heart disease, cancer, stroke) increasing 'Epidemiological polarisation'	Heart disease, cancer, stroke	Heart disease, cancer, stroke (but at later ages)
Age structure	Young	Initially becoming younger Later initial shrinking of child/youth segment	Shrinking child/youth segment Expanding working age and older segments (ageing 'from the bottom')	Ageing continues (increasingly 'from the top')

An ageing world – historical trends

Academic and government attention to population ageing are of comparatively recent origin, most focus in the past having been on issues of demographic growth rather than structure. As will be apparent from the above discussion, though, the process of demographic ageing has a lengthy history, the inevitable accompaniment to fertility decline. France was the first nation to undergo significant structural ageing, reflecting its lead position in the European fertility transition. Sustained fertility decline began in France in the closing years of the eighteenth century and, by 1851, 6.5% of the population was aged 65 and above. Other countries followed suit as fertility transition progressively spread across the continent. In some cases the process was amplified by emigration flows. Sweden, for example, experienced very heavy emigration of young adults between 1860 and 1900, which left the country increasingly peopled by the older generations. Ireland was another case where emigration played an important role.

The same tandem process of fertility decline and ageing occurred in European populations overseas (e.g. USA, Canada, Australia, New Zealand). Migration also entered the equation in these countries, in this case though through the dampening effect of large youthful migrant inflows. In Australia the French mid-nineteenth century 6.5% 65 and over level was reached in the early 1930s, in New Zealand and the USA later in the decade, and in Canada around 1940.

Population ageing was briefly slowed down in many of these countries by baby booms following the Second World War, but picked up again when fertility resumed its downward course. As discussed above, this post-war period has also seen the seeds of ageing sown over much of the developing world with the progressively widespread diffusion of fertility decline.

Table 4.3 details the progress of the ageing transition over the past 50 years and likely trends over the coming 50. Countries have been classified according to the proportion of their population 65 years of age and over: 'young' (fewer than 4%), 'youthful' (4% to 6.9%), 'mature' (7% to 9.9%), 'aged' (10% to 19.9%), 'very aged' (20% to 29.9%) and 'hyperaged' (30% and over). The first four categories follow Cowgill and Holmes' (1970) classification. The latter two are new categories to allow for the greater heights to which ageing will rise over the coming decades.

By this classification there are currently 48 'aged' countries and 14 'mature' ones, all others rating as either 'youthful' or 'young'. Italy (18.1%), Greece (17.5%), Sweden (17.4%), Japan (17.2%) and Belgium (17.0%) presently have the highest proportions of persons aged 65 and over. What precisely transpires with these percentages over the coming years can of course not be foretold as that will depend on trends in fertility, mortality and migration. But unprecedented increases in the proportionate strength of older persons lie ahead

Table 4.3 Ageing transition: number of countries and percentages of population aged 65 and over, 1950–2050

	Under 4% *Young*	*4–6.9%* *Youthful*	*7–9.9%* *Mature*	*10–19.9%* *Aged*	*20–29.9%* *Very aged*	*30% and over* *Hyperaged*
1950	92	58	32	10	0	0
1975	99	47	18	28	0	0
2000	76	54	14	48	0	0
2025	30	48	36	50	28	0
2050	14	23	18	67	58	12

Source: United Nations, Population Division (UNPD) (2003) *World Population Prospects: The 2002 revision population database* (http://esa.un.org/unpp).

for virtually all countries. Taking the United Nations' most recent medium projections as a guide, by mid-century there will be 12 'hyperaged' countries (led by Japan – 36.5%) and another 58 with 'very aged' populations. Significantly, by that stage a number of former developing countries will have joined the pronounced ageing ranks, South Korea and Singapore in fact being projected to have entered the 'hyperaged' category by that date. Figures 4.3 and 4.4 show the present and projected global distributions of the various population age categories.

The ageing transition is also usefully charted in terms of dependency ratios; that is, the proportions of younger people (aged under, say, 15) and older persons (say 65 and over) relative to working-age people. The limitations of dependency ratios for policy development should be noted, however, because, as noted below, people do not automatically become 'dependent' when they reach a given age (60 or 65 and over), especially in increasingly healthy older populations. Nevertheless, it is clear that in pre-transitional populations the ratio tends to be in the order of 80–100 'dependants' per 100 'workers', with almost the total dependency 'burden' in the child ages. As alluded to above, early transition often sees the ratio increase with a swelling of the child component. With the onset of fertility decline, however, the ratio begins to fall away as the larger earlier birth cohorts move (survive) through to working age. After several decades of favourably low dependency (in the 35–50 range) the ratio ultimately rises again as the remaining survivors of the large earlier birth cohorts reach age 65. Today a major feature of almost all developed countries is this shift towards the upper end, with elderly dependency ratios increasing and in a number of cases (e.g. Japan, Italy, Spain, Greece) already outweighing the child dependency component.

The oldest old

By definition, the 60 and over or 65 and over age group covers a very wide range of years and, for both analytical and planning purposes, it is important to distinguish between the *young-old* and the *old-old* (McCracken, 1985; Kinsella and Velkoff, 2001). The major trend from this perspective has been a gradual ageing of the aged, the old-old (80 and over) swelling from one-tenth to one-sixth of the world's older (65 and over) population between 1950 and 2000. Driving this trend has been the upward movement of the large birth cohorts of the late nineteenth and early twentieth centuries and greater proportional improvements in survivorship probabilities for the oldest old over the period. To use Australia as an example, between the late 1940s and the century's end, the probability of newborns surviving to age 80 rose from 29% to 60%, compared with from 70% to 88% for attaining age 65. Apart from a brief minor reversal between 2015 and 2025 due to the arrival of post-Second World War baby boomers into the 65 and over ranks, the old-old will continue to outpace their younger counterparts and by mid-century are likely to comprise more than a quarter of all older persons. Similar data are to be seen in many developed countries. In the United

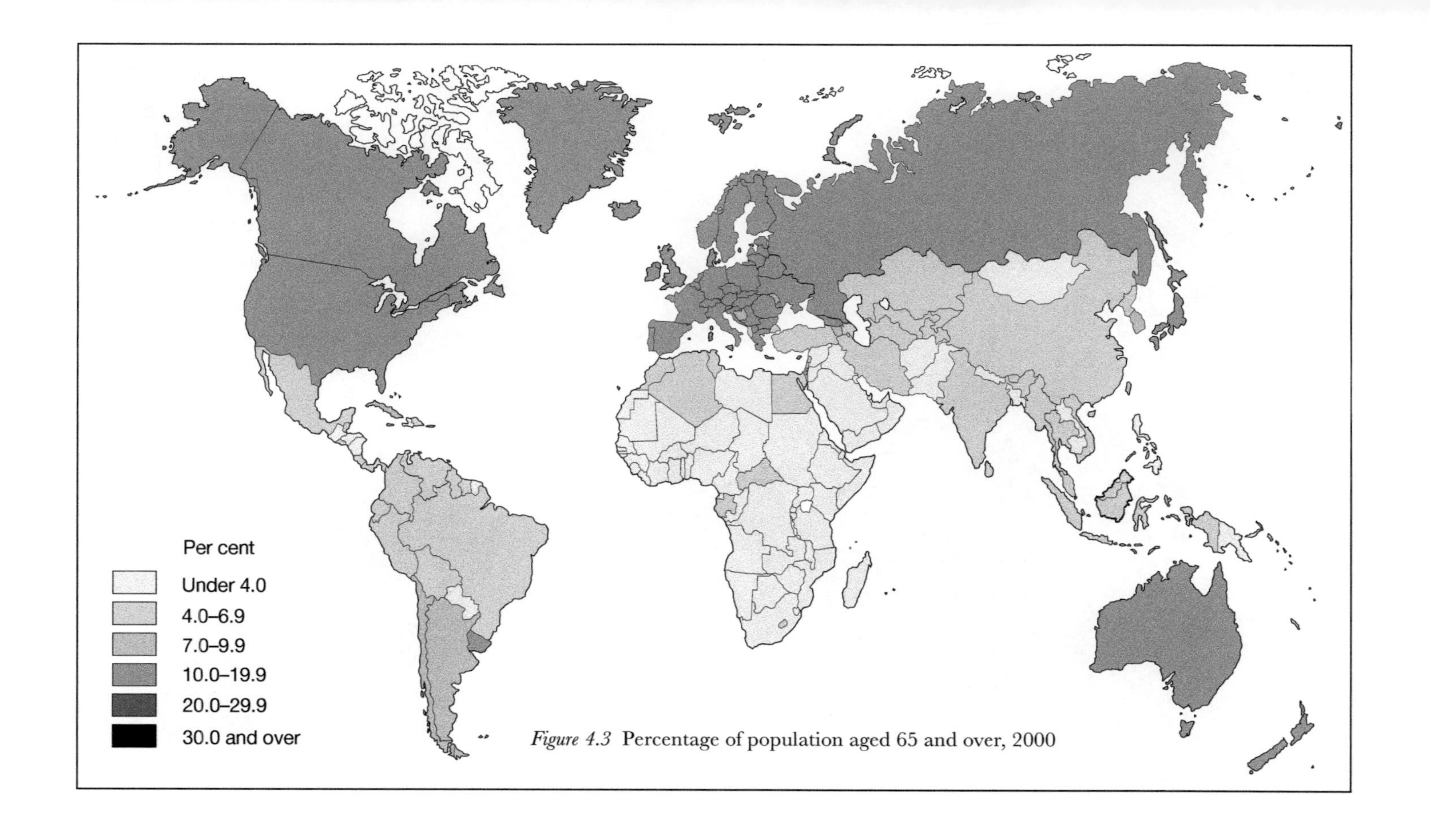

Figure 4.3 Percentage of population aged 65 and over, 2000

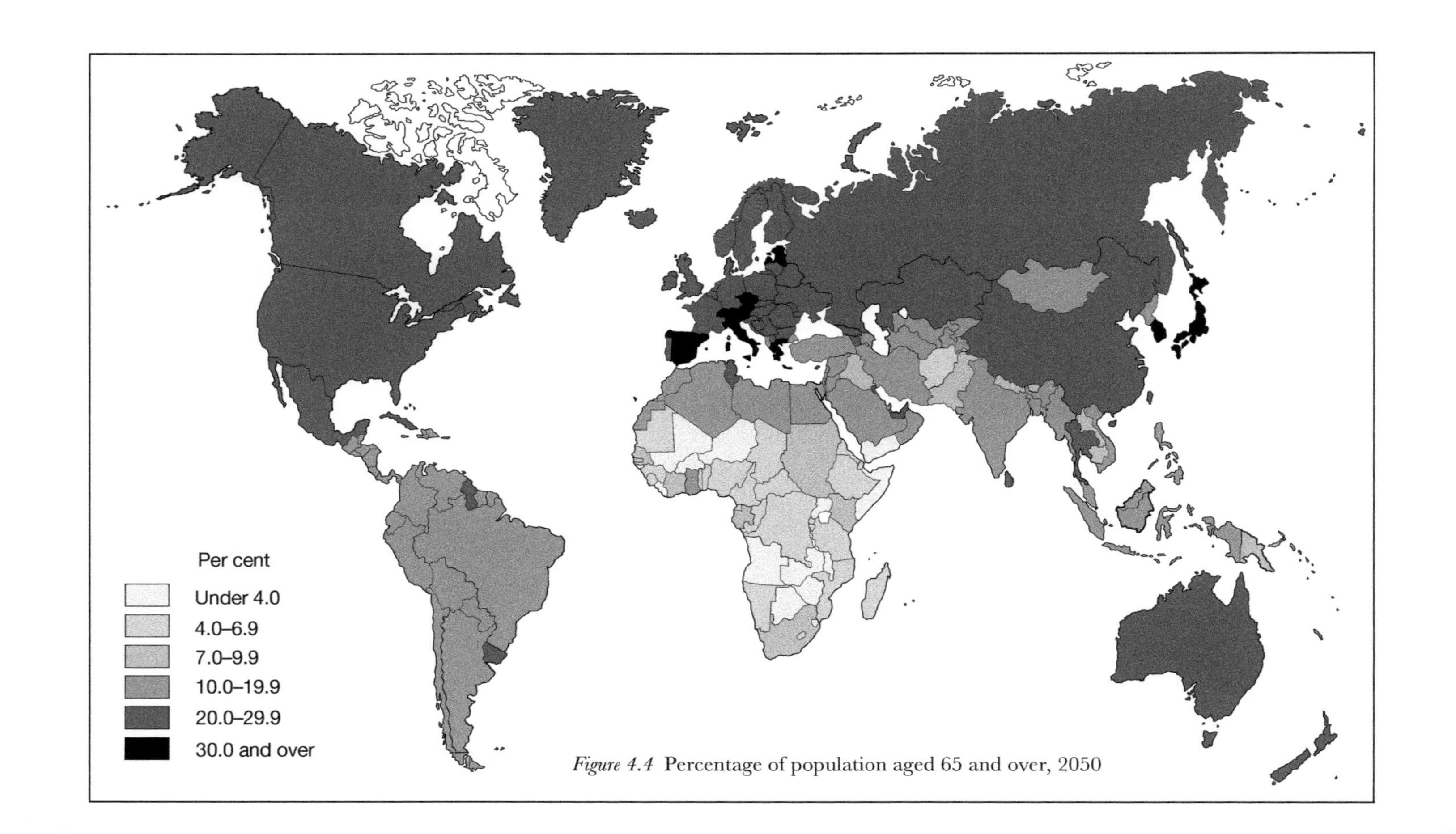

Figure 4.4 Percentage of population aged 65 and over, 2050

Kingdom, for example, the oldest old group are likely to grow in percentage terms from 4% to 8.5% of the population by 2050 (Bartlett and Phillips, 2000).

These developments are in fact seeing the very concept of the 'oldest old' itself changing, and numbers of people of what is currently viewed as extreme old age (say over 95 or 100) will grow considerably (Kinsella and Phillips, 2005). It is at this end of the age spectrum that lifetime benefits of better health and nutrition as well as the technical supports to keep older people living longer appear to be having major effects.

The ageing of the aged population has huge implications for social, medical and economic policies in ageing, as levels of need and dependency tend to escalate in late old age. Even if future cohorts of the oldest old are more healthy than those of today, there will still be considerable increases in demand for services, pensions and personal care assistance. In terms of ageing and place, the spatial distribution of the oldest old people will be very important for planning care and services. Will they mainly be in urban areas, nearer to health and welfare services, or will they be in rural areas? What financial resources will they have? Will their families be living near or will they have migrated? Indeed, the family carers of people in their 80s and 90s will almost always be of near-retirement age themselves, so a double question of care and resources arises. These questions pose huge challenges for policy-makers and are likely to be of considerable importance for applied research in coming years.

The most dramatic rapid gains in longevity have mainly been in East Asia, where life expectancy at birth increased from less than 45 years in 1950 to more than 72 years today and is at or around 80 in Japan, Hong Kong and Singapore, especially for females (Phillips, 2000; Phillips and Chan, 2002). Indeed, a widening of the sex differential in life expectancy has been a central feature of mortality trends in developed countries in the twentieth century, with women generally outliving men by as much as five or six years, although in some high-income countries the gap has narrowed slightly in recent years.

Epidemiological transition (ET)

Underlying the momentous mortality decline populations experience during the course of demographic transition are major changes in causes of death. Progress through the transition sees infectious diseases conquered or receding greatly and chronic, degenerative conditions, sometimes called non-communicable diseases, come to dominate. With this change comes a rising age at death as infectious diseases exact their greatest toll among the young whereas degenerative diseases generally take years to develop and hence primarily cause death at older ages. These changes have become known as the epidemiologic(al) transition, following introduction of the term in a paper by Abdel Omran in 1971. While subjected to some evolutionary critique, Omran's paper has become a classic and, with its various modifications since, has done a great deal to stimulate enquiry (Caldwell, 2001; Gribble and Preston, 1993; Phillips and Verhasselt, 1994).

The demographic, epidemiological and ageing transitions

The original paper envisaged epidemiological transition as a three-stage process, parallelling the demographic transition: an 'Age of Pestilence and Famine', succeeded by an 'Age of Receding Pandemics' and finally an 'Age of Degenerative and Man-Made Diseases'. The suggested finality of the third stage has subsequently been proven premature by continuing epidemiological evolution in developed countries. Mainly unforeseen in the original formulation were the large decline in cardiovascular disease mortality over the past several decades, the related delaying of deaths from the leading degenerative diseases (heart disease, cancer, stroke) to later ages, the rise of 'social pathology' deaths, and the dual emergence of new infectious diseases (such as HIV/AIDS, hepatitis C and perhaps atypical pneumonia/SARS, among others) and resurgence of old ones (such as drug-resistant tuberculosis and malaria). These developments have seen the transition model expanded to incorporate a fourth stage (Figure 4.5). Olshansky and Ault (1986) have coined this new stage the 'Age of Delayed Degenerative Diseases', while Rogers and Hackenberg (1987) have suggested the label 'Hybristic Stage' – to reflect the increasing influence of social pathological and lifestyle-related mortality (such as substance abuse, cirrhosis of the liver, suicide, homicide and accidents). The development of this phase has varied both socially between groups in the population and geographically across regions and within urban areas. Lower socio-economic groups, for instance, have generally not enjoyed such large declines in heart disease and stroke as those of higher status, while hybristic social pathology mortality shows marked racial differentiation in the United States (Table 4.4), Australia and many other developed and middle-income countries.

It is worth also noting that Omran's 'Age of Pestilence and Famine' first stage probably does not do justice to all pre-modern societies – i.e. the hunting and gathering that preceded the shift to sedentism with domesticated animals and plants. As Cohen (1989) has pointed out, the low population numbers and dispersal, mobility and the like would have inhibited person-to-person infections among many prehistoric hunting and gathering groups. Many of the acute interpersonal infectious killers such as influenza, smallpox, measles and diphtheria probably came later with sedentism, from the animals that were domesticated.

Indigenous populations and epidemiological transition

A widespread and especially unfortunate feature of mortality/epidemiological transition has been the disastrous health consequences of European colonisation for indigenous peoples. Wherever it has occurred European colonisation has meant loss of land, marginalisation and cultural and social upheaval for the indigenous populations concerned. The periods of first contact were mortality

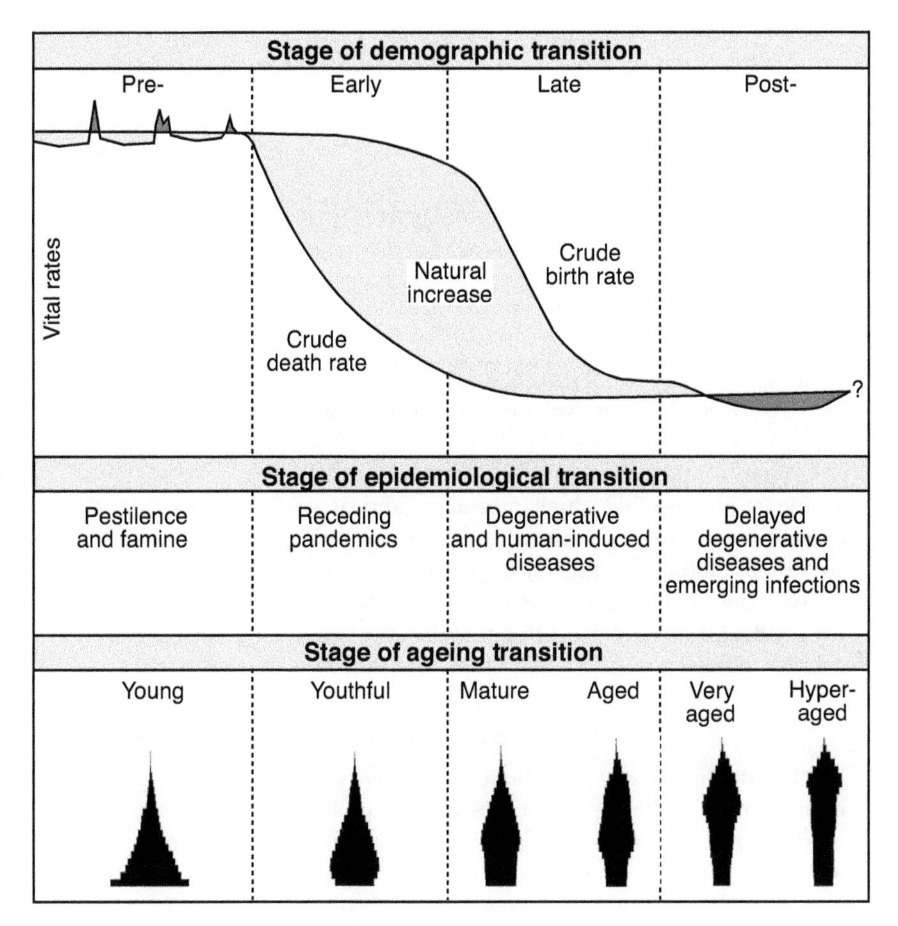

Figure 4.5 The demographic, epidemiological and ageing transitions

disasters for aboriginal inhabitants across the globe – in North America, Australia, New Zealand, southern Africa and many countries of Latin America – in part due to direct armed conflict, but generally more insidiously and savagely through introduced infectious diseases to which the populations had no immunity. From these calamitous beginnings, colonised indigenous peoples the world over have remained disadvantaged in health. The health gap varies considerably, from the seemingly intractable 20-year life expectancy at birth disadvantage indigenous Australians bear in relation to the Australian population as a whole, to the somewhat narrower differentials in the USA (5–6 years), Canada (5–7 years) and New Zealand (7–8 years). However, nowhere have colonised peoples achieved parity of health status with their colonisers (Table 4.5). Marginalisation and social and cultural upheaval, often accompanied by poverty, have produced an epidemiological constellation of multiple disadvantage: high death rates from suicide, assault, heavy drinking,

Table 4.4 Deaths due to hybristic causes, by race, United States, 2000

	White (all persons) %	*Black (all persons) %*	*Black (males) (25–34 years) %*	*American Indian (all (persons) %*
Accidents	4.0	4.3	19.9	11.9
Assault (homicide)	0.4	2.8	28.3	1.8
Suicide	1.3	0.7	6.0	2.6
Chronic liver disease and cirrhosis	1.1	1.0	na	4.7
HIV	0.3	2.7	12.4	0.5

Source: Anderson, R. N. (2002) *Deaths: Leading Causes for 2000*, National Vital Statistics Reports, 50 (16). Hyattsville, Maryland: National Center for Health Statistics.

accidents and the like; high mortality (and at younger ages than the non-indigenous population) from heart disease and several other degenerative conditions; and alarmingly high rates of diabetes. One in six Aboriginal deaths in Australia in 2001, for instance, was due to external causes, while a further 7% resulted from diabetes. On top of this, the community's death rate from circulatory diseases was twice that of the total Australian population.

Life-expectancy reversals

Worldwide, practically all nations over recent decades have shown continued improvement in life expectancy, especially at older ages. Even the indigenous mortality rates noted above have tended to show some improvements. Yet there have been some exceptions. For example, in some of the former Soviet Union republics the economic collapse, social disruption and political upheaval following the dissolution of the USSR caused death rates to rise sharply, particularly among males. In the Russian Federation, for instance, male life expectancy at birth fell below 60 years and remains very low by developed-nation standards. Even more catastrophic falls in life expectancy have occurred in a number of African countries due to the HIV/AIDS epidemic. Zambian life expectancy at birth, for instance, is estimated by the UN to have fallen from around 51 years in the mid-1980s to 34 at the end of the century, Zimbabwe's from 60 to 37 and Botswana's from 63 to 48. And significant further declines will occur before the epidemic is played out.

These life-expectancy reversals serve as further reminders of the many variants of epidemiological change and the falsity of any notion of seemingly inevitable improvement and progression through the stages. The African AIDS crisis is a particularly tragic illustration, highlighting that sharp counter-transition back to an earlier epidemiological age can unfortunately occur.

While such things as AIDS and drug-resistant tuberculosis are dramatic reminders that the pronouncements of the 1960s that the era of epidemic disease in developed nations was coming to an end were seriously premature, we

Table 4.5 The indigenous mortality burden – some international comparisons

	Life expectancy at birth (years)		*Life expectancy at age 65 (years)*		*Infant mortality rate (per 1,000 live births)*
	Males	*Females*	*Males*	*Females*	
Australia 1999–2001					
Aboriginal and Torres Strait Islander population	56.3	62.8	8.0	9.9	12.8
Total population	77.0	82.4	17.2	20.7	5.4
Canada 2000					
Registered Indian population	68.9	76.6	na	na	4.0[1]
Total population	76.3	81.8	15.9	19.8	3.7
New Zealand 1995–7					
Maori population	67.2	71.6	12.2	14.5	7.9
Total population	74.3	79.6	15.5	19.0	5.7
United States 1997–9					
American Indian and Alaskan native population	67.4[2]	74.2[2]	na	na	9.1
Total population	73.8	79.4	16.0	19.2	7.1

Sources:

Australian Bureau of Statistics (2002) *Deaths Australia, 2001.* Canberra: ABS.

British Columbia. Provincial Health Officer (2002) *Report of the Health of British Columbians. Provincial Health Officer's Annual Report 2001. The Health and Well-being of Aboriginal People in British Columbia,* Victoria, BC: Ministry of Health Planning.

Indian and Northern Affairs Canada (2002) *Basic Departmental Data, 2001.* Ottawa: Department of Indian Affairs and Northern Development.

Statistics New Zealand (1998) *New Zealand Life Tables, 1995–7,* Wellington: Statistics New Zealand.

US Department of Health and Human Services (2002) *Health, United States, 2002,* Hyattsville, Maryland: National Center for Health Statistics.

Notes

1 British Columbia, Status Indians

2 1996–8

also need to be careful to not underestimate the continuing contribution of less spectacular infections to present-day mortality. The International Classification of Causes of Death (ICD) encourages this underestimation. Over its various revisions the first chapter of the Classification has been 'infectious and parasitic disesases' and commentators have frequently taken deaths in that chapter (now down to close to 1% of total deaths in most developed nations) as representing virtual elimination of infectious diseases as killers. Buried away in the classification, though, are other deaths with infectious aetiologies: those directly identified, such as influenza and pneumonia, but also numerous others – e.g. liver cancer deaths resulting from hepatitis B and C, cervical cancer deaths due to human papiloma virus (HPV), some genito-urinary conditions, rheumatic heart disease, and the like. A recent New Zealand study, for instance, suggested that around 7% was closer to the true infectious disease death toll (Christie and Tobias, 1998). Apart from this hidden side of infectious disease there is the 'associated' role infectious conditions such as influenza, pneumonia and septicaemia play in many deaths. As an illustration, in Australia in 2001, influenza and pneumonia were recorded as the *underlying cause* in 2702 deaths (2.1% of total deaths). The two infections, though, were also *associated* with 3046 cancer deaths, 1961 cerebrovascular disease deaths, 1858 chronic lower respiratory disease deaths, 1430 ischaemic heart disease deaths, 1377 disease of the nervous system deaths, and many assorted others.

Ageing and the later stages of epidemiological and demographic transition

The delayed degenerative diseases stage of epidemiological transition is of great importance for ageing populations as its characteristics will greatly influence the levels of disability, healthy life expectancy and needs for health and social care, as well as the ability of people to live alone and of their families to care for them. Indeed, as populations age demographically, their epidemiological profiles are also likely to become more varied and there is evidence that some older persons, probably for a mixture of lifestyle and genetic factors, may live long and healthy lives, well into their 80s, while others may experience the 'creaking door' syndrome of longer life but gradually deteriorating health.

While 'late-stage' epidemiological transition countries all share degenerative disease-dominated mortality profiles, important variations occur beneath that broad commonality. Table 4.6, for example, shows the quite different sources of mortality, which is similar overall, between three ageing countries, although as these are based on crude death rates the usual cautions should be noted. The most marked differences are the considerably lower percentage of deaths from heart diseases seen in Japan than in Australia and the USA and, conversely, the Japanese population's higher percentage of cerebrovascular diseases.

It seems clear from recent data that even the low- and middle-income countries are moving steadily towards a growing burden of disease from chronic and degenerative conditions, sometimes called non-communicable diseases. As Figure 4.6 shows, this is especially the case for the older age groups in such countries, although non-communicable conditions begin to be very important contributors to the burden of disease from middle age onwards. This stage of transition can be especially problematic for the countries concerned as they are faced with a situation of epidemiological polarisation; that is, co-existing old style (infectious) and new style (degenerative) disease burdens. The former tend to be concentrated among the poorer section of the population, while the latter are more associated with the higher social classes who in turn are generally more powerfully placed to win the competition for scarce health resources. However, we should note that some older persons everywhere remain vulnerable to acute infectious conditions, especially pneumonia-type ailments. New emerging infectious disease variants can also be risky for older persons. For example, the outbreak in many countries in 2003 of SARS (severe acute respiratory syndrome) affected many older persons. In Hong Kong, of the reported 1755 SARS

Table 4.6 Epidemiological variability in late-stage transition countries

Three leading causes of death	*USA 2000 %*	*Australia 2001 %*	*Japan 2000 %*
Diseases of heart	29.6	26.2	15.6
Malignant neoplasms (cancers)	23.0	28.6	30.7
Cerebrovascular disease (stroke)	7.0	9.4	13.8
Cumulative percentage of total deaths	59.6	64.2	60.1

Sources:
Anderson, R. N. (2002) *Deaths: Leading causes for 2000*, National Vital Statistics Reports, 50 (16), Hyattsville, Maryland, National Center for Health Statistics.
Australian Bureau of Statistics (2002) *Causes of Death Australia, 2001*, Canberra: ABS.
Ministry of Public Management, Home Affairs, Posts and Telecommunications (2003) *Japan Statistical Yearbook 2003* (http://www.stat.go.jp/english/data/nenkan/index.htm).

cases in 2003, 18.5% were persons aged 65 and over. More importantly, the death rate was much higher in the older than younger age groups. 188 of the 300 SARS deaths (62.7%) in Hong Kong were among persons aged 65 and over, with the highest mortality rate being among the oldest old (Hong Kong Government, 2003).

Health advances and the health transition

Omran's descriptive model of epidemiological transition has certainly been very useful and has stimulated debate and research. One aspect that has attracted attention is the reasons for improving or changing health. Some researchers suggest placing emphasis on scientific advances as driving many social and other developments, as well as being intricately bound up with modernisation itself. Caldwell argues that 'expressing the change as being essentially socioeconomic subtly downgrades the specific contributions made by public health interventions and especially by breakthroughs in medical science' (2001: 160). Others note that the bulk of early mortality decline from infectious diseases in the West in particular had actually started before the infectious disease immunisation discoveries (McKeown, 1976; McCracken and Curson, 2000). By contrast, it could be said that public health campaigns and in particular the availability of immunisations for childhood killers like measles and others such as tuberculosis have had a greater impact in today's Third World countries although, again, it is difficult to disentangle their impacts from those of general improvements in living standards. Thus epidemiological transition is very varied and, indeed, there may be as many models of ET as there are societies. As well, the transition varies socially and spatially within countries (Caldwell, 2001; Phillips, 1993). Any idea of a single, pre-determined epidemiological pathway as a natural destiny of societies is thus naive and misinformative (Carolina and Gustavo, 2003).

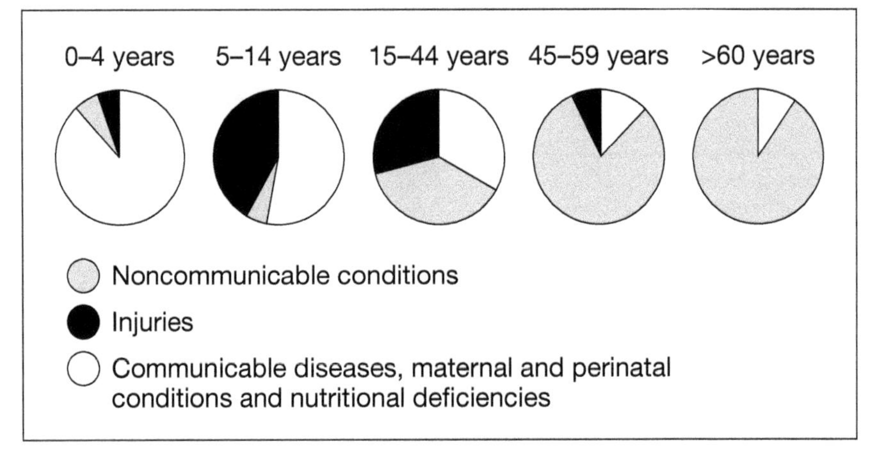

Figure 4.6 Leading causes of the burden of disease by age, both sexes, 1998, low and middle income countries. Source: WHO (2002) p. 15, Fig. 6.

Some researchers outline a broader *health transition* rather than an epidemiological (mortality) transition, comprising a set of interlinked changes involving the demographic movement from high to low fertility, steady expansion of life expectancy at birth and also at older ages, and shifting causes of ill health (morbidity) as well as mortality. The health transition has undoubtedly occurred, albeit in a spatially varied form, and has been fostered by many changes related to modernisation and urbanisation over the past decades, importantly, but not solely, by improvements in health care as well as standards of living and education. However, it is important to remember that health transition improvements can also be lost due to socio-political, economic and other upheavals. The declining health status and reductions in life expectancy in Russia during and after the breakup of the Soviet Union around the late 1980s and into the 1990s and the reversals in the African AIDS-belt nations discussed earlier are telling examples.

Healthy life expectancy: a major question in ageing populations

Whatever the specific mix of underlying reasons, epidemiological transition and a broader health transition have certainly occurred almost everywhere. As well as the radical changes in fertility discussed above, all countries have witnessed a greater or lesser change in the balance of causes of deaths (mortality) and ill health (morbidity) in their populations over the course of the twentieth century and especially in its closing few decades. While mortality has traditionally been used as an indicator of population health status and the need for health and welfare services, it is increasingly recognised that a people's lifetime health status (how healthy or ill they have been) is the major influence on well-being and the need for health and other support services. In low-mortality countries in particular, counting most countries of the West and many in Asia and Latin America, death often follows only as one event after a life which has had a greater or lesser amount of chronic and acute illness (Ratzan *et al.*, 2000). Considerable numbers of people everywhere live for a long time with chronic health problems that do not lead to death but which certainly affect their economic prosperity. For older persons, in particular, this often affects their quality of life and their ability to live independently and conduct activities of daily living (ADLs). The balance between increased years of life and healthy and less-healthy old age is increasingly regarded as a key to understanding the needs and impacts of ageing populations. How this is occurring in different places and between different groups is extremely important.

These issues have seen the traditional concept and measurement of life expectancy extended in recent years to the notion of healthy life expectancy (HALE), combining information on mortality and non-fatal health outcomes. This is a more informative measure of health status, especially in relation to the older population. It is usually reported in official statistics as HALE at age 60. From Table 4.7 (overleaf), it is clear that the wealthier countries have a marked advantage in years of healthy life expectancy and also in greater equality between males and females. In many such countries, males aged 60 may expect to live on average a further 20 years or so, of which around 16 years might be expected to be healthy. By contrast, in countries in the poorest groups such as Zambia, Malawi and Sierra Leone, men may expect to live a further 13 years at 60, but for barely half that time will they be healthy. More than simple longevity, HALE has huge implications for growth of demand for health and long-term care services in the various countries and regions within countries. Health, ageing and place are thus intricately interwoven.

As well as increases in physical impairment and handicap in older age, there is increasing interest in the incidence and prevalence of vascular dementia and Alzheimer's disease as populations age. Based on an extensive review of studies of the epidemiology of dementia, Suh and Shah (2001) have suggested there is perhaps an explanatory model of transition from low-incidence/high-mortality societies to high-incidence/low-mortality societies, in terms of Alzheimer's disease. Other studies point to prevalence rates also increasing for dementia with age. Many studies have noted a trend for prevalence to increase exponentially with age and Jorm *et al.* (1987) found the rate doubled with every 5.1 years of age (from a statistical integration of data from 22 studies). The baseline prevalence figures they found were:

Age group	60–4	65–9	70–4	75–9	80–4	85 and over
Prevalence rate (%)	0.72	1.42	2.82	5.60	11.11	23.60

These data are in line with other studies that indicate the high prevalence of dementia comes mainly in the 85 and over age groups. So, when the numbers of people in the oldest-old groups increase as indicated earlier, there is likely to be a much greater demand for dementia care services and associated support

Table 4.7 Life expectancy and healthy life expectancy (HALE) in years

	Males			*Females*		
	Life expectancy at birth	*Life expectancy at 60*	*Healthy life expectancy at 60*	*Life expectancy at birth*	*Life expectancy at 60*	*Healthy life expectancy at 60*
	2000[1]	*2000*[1]	*2001*[2]	*2000*[1]	*2000*[1]	*2001*[2]
High life expectancy						
Japan	77.5	21.3	17.1	84.7	26.8	20.7
Sweden	77.3	20.6	16.5	82.0	24.3	18.5
United Kingdom	74.8	18.8	15.0	79.9	22.7	16.9
United States of America	73.9	19.6	14.9	79.5	23.1	16.6
Medium life expectancy						
Brazil	64.5	16.2	9.4	71.9	19.6	13.0
Egypt	65.4	15.1	9.4	69.1	17.3	9.2
Russian Federation	59.4	13.3	8.5	72.0	18.3	12.7
Bangladesh	60.4	14.7	9.4	60.8	15.7	10.9
Low life expectancy						
Afghanistan	44.2	13.5	4.9	45.1	14.7	8.7
Zambia	39.2	13.4	7.5	39.5	15.5	10.0
Malawi	37.1	12.7	7.2	37.8	14.8	9.5
Sierra Leone	37.0	12.9	5.5	38.8	14.3	8.5

Sources:

1 World Health Organization, *Life Tables for 191 Countries: World mortality in 2000* (http:// www3. who.int/whosis/).

2 World Health Organization (2002), *The World Health Report 2002*, Geneva: WHO (Annex Table 4).

for a condition that can be very difficult to deal with for family and health services. These data also remind us that healthy life expectancy relates not only to physical but also to psycho-social and psychological well-being.

A paramount research area of epidemiological transition for ageing populations will be the way in which the transition subtly varies as older populations themselves age. Given the HALE data above, we might reasonably expect a healthier younger-old population, for a range of reasons, including better lifetime health care, nutrition, education and living standards. However, the accompanying trends in later onset disability and whether there is any compression of morbidity will very much determine the health status and ability to live alone of the oldest-old population in particular. These factors will also strongly influence the costs of health and long-term care. Many economists and politicians use alarmist language about a 'burden of ageing' but, if future older generations are healthier and economically better prepared for retirement than those of today, perhaps this burden will not be as heavy as some fear (Phillips, 2000). However, in the middle-income countries and especially those in the Asia–Pacific and parts of Latin America, where recent demographic ageing has been quite rapid, the lack of formal care and social trends that still emphasise family care may place considerable burdens on the family in the coming decades (ESCAP, 2002).

Compression of morbidity?

A major debate in the discussion of changing life expectancy and the incidence and prevalence of morbidity and disability at older ages relates to the amount and duration of ill health people will experience in old age. One suggestion is based on the compression-of-morbidity hypothesis dating from the early 1980s when Fries (1980) and others postulated that in future chronic disease would probably only occupy a small part of the entire lifespan and would be compressed into a short while at the end of life, which was envisaged to be about 85 years (in future it is clearly likely to be even higher). However, a second scenario suggests that life expectancy will increase, perhaps slowing as the genetic life potential of the species is approached, but that it will be accompanied by longer periods of ill health and disability. Estimates of the future 'burdens of illness' from demographic ageing are very dependent on which of these two contrasting scenarios may be more accurate, compression of morbidity or longer life and possible worsening health. Unfortunately, detailed data internationally are too scanty to give definitive support to either scenario. Some recent studies have identified the potential for compression of morbidity, especially with regard to health care expenditure which is increasingly focused in a short while before death (Kalache and Keller, 2000). Others point to indications of continuing chronic poor health and disability being widespread among high proportions of some older populations.

Ageing populations and economic costs

We certainly do not want to add to the moral panic that is widely associated with socio-political and economic analyses of demographic ageing. Two main areas related to costs of older populations tend to stand out: the so-called 'pensions gap' and the 'spiraling costs of health and social care'. The debate on pensions is technical and often relates to public versus private systems and fully funded systems versus those that 'pay as you go', with the younger generation paying for the current retired population's pensions. There are vast international differences in pensions and retirement income provision and it is very difficult to extrapolate findings from one country to another because of the different base levels of provision, income, taxation and other benefits in kind that different systems offer (see, for example, various OECD studies such as OECD 2000; Casey and Yamada, 2002; Yamada, 2002). Currently, many of the major European countries such as France and Germany face serious problems in funding their public pensions schemes. The United Kingdom is moving increasingly towards private-sector provision with a residual state pension and the United States of course is very advanced in private-sector provision. Most low- and middle-income countries have only very small proportions of their citizens covered by any form of formal pensions and family care; individual savings or continued work remain the main sources of income for older persons.

In many developed nations, government concerns regarding the pension burden are seeing policy attention being given to encouraging employment until older ages. The United Kingdom, for example, proposed in 2003 outlawing ageism in employment advertising and banning employers from setting a retirement age below 70. Internationally, the attraction of early retirement, discrimination against older workers by employers and poor work incentives in pension schemes have produced significant declines in older workers' activity rates over the past 15 to 20 years, but the public costs of this lost workforce potential are now being increasingly recognised. To this end, a wide sweep of countries over recent years have reformed their pension systems to reduce pension outlays and promote employment of older workers. These changes have ranged from raising pension eligibility ages, lifting pension ages for women to equalise them with those of men (usually age 65), discouraging early retirement and encouraging work after the normal pension age by offering higher payments upon deferred pension uptake.

In terms of health costs and the economic impact of population ageing, in countries for which solid data exits it is difficult to escape the fact that per capita health care costs do appear to be strongly age-dependent. For example, the Australian government's per capita outlays on health by age group in 1998–9 were estimated to be A$1979 (50–9), A$3317 (60–4), A$4587 (65–9), A$6840 (70–4) and A$12,786 (75 and over), clearly indicating an escalation of costs particularly in the oldest age group (Guest and McDonald, 2002). Netherlands data meanwhile indicate per capita health care costs to increase exponentially from age 55 onwards (Polder *et al.*, 2002). However, internationally, it is very difficult to pronounce definitively on the subject of age and health care costs, as expenditure on older persons' health care is strongly influenced by the funding nature of nations' health services (whether, for example, they depend on individual payments, insurance or collective coverage, all of which clearly affect their use by retired persons of limited means in particular).

Nevertheless, when the costs of social care and long-term care for older persons are included, it is very difficult to avoid the conclusion that, even if today's older populations are on average healthier than in the past and if in the future they are likely to be yet healthier, the fact of increasing numbers reaching oldest-old age will inevitably impact on costs. Reviews of studies have shown that rates of impairment, disability and handicap, and quality of life, are strongly age-related (Barbotte *et al.*, 2001). Yet almost all studies acknowledge that the impact of declining health and increasing disability is also very much influenced by socio-economic factors, notably the breadth and coverage of health and social care systems. Thus older populations should not always automatically be blamed for increasing health care expenditures as higher costs of acute care and health services utilisation are often related to the closeness of death, not always to age. Also, rising health care costs are not attributable solely to population ageing. Other factors such as new medical technology, rising consumer demand and escalating prices are major expenditure drivers.

Lifespan perspectives, active ageing and successful ageing

In recent years, there has been increasing recognition of the importance of viewing people's health and well-being, and also their social and economic activities, in a life course perspective rather than compartmentalising 'the elderly' into a separate category. This is now widely adopted by more enlightened policies which view ageing as part of the life course and emphasise the inter-generational relations that are crucial at the family and societal levels. For future policies and research, this perspective will no doubt become increasingly important. The socio-spatial variations in health and development over people's life courses will be extremely important to the need for differential levels of social support, formal and informal.

The WHO has developed an associated policy of *active ageing*, which it sees as 'the process of optimising opportunities for health, participation and security in order to enhance the quality of life as people age' (WHO, 2002: 12). The term 'active' refers to continued participation in social, economic, cultural, spiritual and civic affairs, not just the ability to participate in work. Older people who retire can be just as active as many others and can contribute to families and communities. Many feel that active ageing has a cumulative effect and extends healthy life expectancy, providing double benefits to individuals and society. Furthermore, people with disabilities and handicaps are also encouraged in the active ageing framework to participate as much as they wish or can.

A related concept is *successful ageing*, which is related to the broad issues of coping and adaptation in later life (Rowe and Kahn, 1998). There are growing numbers of older persons who do not exhibit chronic health problems and declining cognitive skills that many have previously assumed are almost an inevitable part of the ageing process. The process of adaptation has been seen as a key component to successful ageing and, as illustrated in the Berlin Aging Study, older adults can compensate for losses and declines and also have the potential for further growth and development (Baltes and Mayer, 1999). Successful ageing can be thought of partly as successful psychological adjustment to ageing and as maximising desired outcomes and minimising undesired ones such as losses. It has been explained in part as a process of selective optimisation with compensation to manage one's life and has been conceptualised as a hierarchy, consisting of three tasks shown in Figure 4.7. There is discussion as to how those who age successfully differ from others and also the extent to which external factors have a role in the process. Nevertheless, the concept clearly has many implications for designing policies and interventions that can help set the scene for successful ageing and minimise factors that can inhibit it. That successful

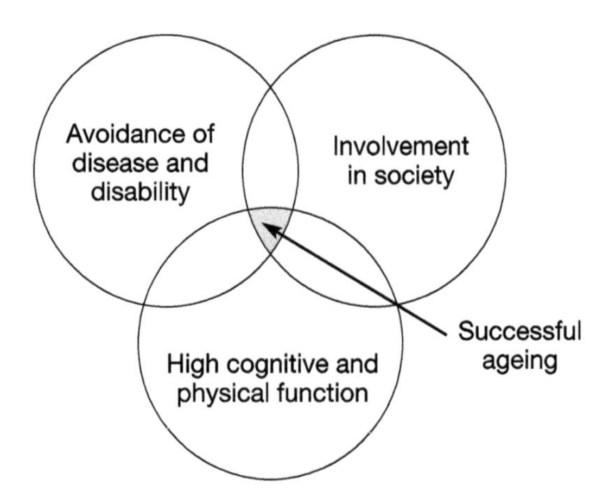

Figure 4.7 Successful ageing (source: Rowe and Kahn, 1998)

ageing varies spatially and socially is almost certain, but little systematic, comparative research has been conducted, so this is a major area for future studies.

Note

1 The use of these ages is simply one of operational convenience, and can be traced back to historically determined retirement and pension eligibility ages in numerous Western nations, rather than there being any sense that they genuinely represent the beginning of 'old age'. The relevance of particular chronological markers also varies considerably between societies. For instance, in those less developed nations where life expectancy is very low and persons over 60 or 65 are much fewer, a lower old-age threshold could be more appropriate. No chronological marker, however, can really capture the functional and socially constructed meanings of age.

5

RESIDENTIAL HOMES

From distributions in space to the elements of place

Gavin J. Andrews

Residential homes for older people have been a focus of sustained research interest and debate. This is because of their prominence in the provision of long-term care for older people in many countries, and also because of their unique and sometimes controversial role as semi-institutional environments which fulfil diverse accommodation and care functions, to some extent filling the accommodation and care 'gap' that lies between fully institutional long-term health care and supported home dwelling. Indeed, residential homes are distinct from nursing homes or long-stay hospital wards because their care function only extends to personal care. However, like these other forms of care, they do provide accommodation and full board. Research interest in residential homes has occurred across a number of academic sub-disciplines including medical/health geography, medical sociology, environmental psychology, social policy and health economics, as well as across health professional research disciplines including nursing, occupational therapy and social work research (Andrews and Phillips, 2002). With a particular emphasis on geographical studies, and other studies that contribute to knowledge of residential homes as complex places, this chapter traces the broadening scope of empirical foci, changing perspectives and theoretical development on the subject over a 20-year period.

The literature on residential homes is complicated because the regulatory definitions of what constitutes a residential home, and its particular social and health care functions, differ both between countries and over time. Additionally, even when definitions are consistent between countries, the character of a residential home may be country-specific. In the UK, for example, residential homes are typically small businesses, owned and managed by a husband-and-wife partnership, based in an Edwardian or Victorian building and accommodating 10–20 residents (Andrews and Phillips, 2002). By contrast, in much of North America, residential homes are considerably larger, are owned and managed by a large companies or hospitals and visually resemble apartment complexes. The above examples only take stock of the most basic physical and ownership characteristics. Different management, regulation and care regimes further complicate residential care as a discrete category. To provide a sharper perspective, in this chapter I concentrate mainly on the UK residential sector and research on it. Certain debates and concepts may, however, be transferable to residential homes and residential homes research in other countries. In terms of research perspectives, the chapter exposes a gradual movement from a research tradition dominated by macro-scale (regional and national) perspectives on the distributive and allocative features of the residential sector (in 'space'), to a tradition that now also includes micro-scale perspectives on care, management and home life 'within' residential homes (in 'place', understood as a physical, symbolic and social construction). The first categories of research discussed (historical perspectives; underlying demand factors; policy, funding and levels of provision; distributive features in spacc; regulatory features) largely relate to these macro-scale perspectives, while the final categories (residents, ownership, staffing) largely relate to these micro-scale perspectives. However, much literature and many debates and issues do cross-cut these categories. The final discussion signposts some areas for future research enquiry, both in terms of empirical subjects and the ways they might be studied. These are shown to cross-cut some of the traditional sub-disciplinary strata of human geography and also engage substantively with place, image and identity.

A UK historical perspective

It is widely regarded that the first dedicated and substantive study of the UK residential sector is Townsend's *The Last Refuge* (1962). Pioneering research in the field, *The Last Refuge* described the long history and development of residential care in the UK. Its particular strength is in describing the changing perceptions in British society of different forms of communal living and changing policy orientations. Beginning his historical journey in the eighteenth century, Townsend argued that communal living was rare because of lower life expectancy, higher levels of family cohesion and responsibility and a lack of interest and responsibility for provision by either the state or private companies. Townsend suggested, however, that, by the early twentieth century, combinations of rising life expectancy, the loosening of family ties and an expanded working-class population created a demand for long-term health care facilities. Townsend reported that initially most new supply was in the form of state-provided geriatric hospitals which opened for the working classes. In addition, a small number of private 'homes for gentle folks' opened for the middle classes (which resembled hotels) and a very small number of charity homes supplemented this provision. Indeed, the concurrent uneven split, between a minority of homes being privately and voluntary- or charity-owned and managed and the majority of homes being public- or state-owned and managed, formed the basic structure of the national residential care sector in the UK for many decades thereafter.

As Townsend described, better organisation of public residential services occurred after the Second World War with the inception of the British NHS and the introduction of a much-expanded welfare state. At this time, state support for residential care increased dramatically under the responsibility of local government. A small private sector still catered for privately paying clients and was able to grow itself, as Townsend suggested, largely due to greater affluence in society and because of the negative image of local-authority provision, which was arguably justified. Townsend did allude to some basic geographical features of early provision, stating that private homes were most frequently located in affluent districts and local-authority homes in poorer districts – an obvious observation perhaps on supply meeting demand geographically, and certainly one that was not formally measured by Townsend. Since *The Last Refuge*, many studies have traced developments in residential provision and, more specifically, the gradual emergence of the private sector as the dominant provider of care in the UK. Indeed, the first phase of substantial growth in the private sector occurred in the 1970s following the closure of many long-stay hospital wards and as public expenditure on local-authority homes could not keep pace with demand. As will be described in the following sections, the second phase of growth occurred in the 1980s following substantive policy and funding changes, while a third phase of stabilisation and localised decline occurred in the late 1990s following further funding changes under the care-in-the-community reforms. Currently the impact on provision of stricter regulatory standards introduced at the start of the new millennium is yet to be determined.

Underlying demand factors

Regardless of any policy or financing orientations that may effect the supply of and demand for residential homes, over time they have still become a common and to some extent necessary feature in society. A combination of two factors, first identified by Townsend, seems to be a highly plausible explanation for this. First, a significant social change underpinning the growth in all communal living environments has been the loosening of family ties and responsibilities. In this regard, Finch and Mason (1993) and Finch (1989) have considered changing family life and kinship relationships. Their research has suggested that responsibility and obligation for caring fully for dependent elders is not longer as automatic as it may have been for past generations, involves different parameters and negotiations and has dramatic consequences for places of care. Hence, while many counties have witnessed policy and service trends towards home and community care (which may well be considered as 'alternative' provision to residential care), for many older people, as soon as being supported in their own home is no longer possible or appropriate, in the absence of a move into a family member's home, few care and accommodation options are available to them other than forms of communal living.

Second, the demographic phenomenon of population ageing plays an important role as an underlying demand factor. Due to decreased fertility and increasing life expectancy, the proportion of older people within the UK population has long been steadily rising. For example, the proportion of people of pensionable age in the UK grew by 30% between 1951 and 1971 (OPCS, 1985). These trends look set to impact even more heavily in the future. As Bartlett and Phillips (1995) have told us, during the next 25 years the UK population will increase modestly from 58 million to about 62 million. Thereafter, over the following 50 years, it is likely to reduce to about 57 million. However, between 2011 and 2041 those aged 65 to 79 are expected to increase from 12% to 17% of the population while those aged 80 and over are expected to increase from 4% to 9% of the population (Bartlett and Phillips, 1995). Notably, these figures suggest that not only will the population age, but the proportions of the very oldest people will increase most rapidly. Although it is difficult to assess the extent to which overall disability and dependency levels change nationally as a result of an ageing population, it is probably fair to suggest that, on an individual level, with increasing longevity comes an increase in problems with vision, hearing, mobility, incontinence and dementia, and therefore a corresponding increase in the demand for long-term accommodation and care (Andrews and Phillips, 2002). Indeed, Phillips *et al.* (1988) identified that while only 5–6% of those aged 65 and over live in residential homes or hospitals, 21% of those aged 85 and over do so. Given these forecasts, arguably a substantial amount of residential support will be required for a future ageing population.

Policy, funding and levels of provision

Beyond these underlying demand factors, research has traced the long-standing role that the British state has had in funding residential care and its concurrent impact upon levels of provision and the character of the sector. As a number of studies have told us, on the national level, successive governments have encouraged a greater role for the private sector in residential provision (Smith and Ford, 1998; Phillips *et al.*, 1988; Andrews and Phillips, 2000; 2002). More specifically, as suggested, a policy of closing long-stay hospital wards added to the numbers of potential clients for residential homes and effectively increased the size of the potential market (Bartlett and Phillips, 1996; 2000). However, the catalyst to what was effectively a 'boom' in private provision was the introduction of what was in effect guaranteed state support for residents in private residential homes. Indeed, from November 1983 to 1993, anyone with less than a specified amount of savings qualified automatically for full state benefits which would meet private-sector home fees. This encouraged expansion in the sector by dramatically increasing the number of potential clients and hence reducing the financial risk involved in running residential homes (Phillips and Vincent, 1986a; Andrews and Phillips, 2000; 2002). More generally, residential homes became known for their financial security and profitability. Two other factors also promoted growth in the sector, and were further incentives to starting such businesses. First, in many locations, residential homes were exempt from local property taxes (then called rates) (Phillips *et al.*, 1987a; Andrews and Phillips, 2000; 2002). Second, during much of the 1980s the property market was healthy and house prices were increasing rapidly. The capital gains potential thus provided an added attraction to owning a residential home (Phillips and Vincent, 1986a; Andrews and Phillips, 2000; 2002).

Geographical research has demonstrated that, at the local scale, a number of factors promoted particularly rapid growth in the south of England, and in particularly in coastal regions. As Andrews and Phillips (2000; 2002) outline, first, retirement migration to seaside resorts led to higher numbers of potential clients in these regions (Law and Warnes, 1976; Warnes and Law, 1984; Warnes and Ford, 1992). However, this is not to say that residential homes were directly responsible for this inward migration. Rather, Phillips *et al.* (1988) identified a 'secondary migration model' whereby, on retirement and while living independently, older people move to favoured retirement locations, and then subsequently require residential services as their deteriorating health demands. Second, research has also demonstrated that in some localities planning controls were relaxed in order to encourage growth in the residential sector and provide an opportunity for residential homes to expand or to convert other premises (Phillips *et al.*, 1987a; 1987b; Andrews and Phillips, 2000; 2002). Third, a decline in domestic tourism in some coastal regions left a stock of former hotels and guesthouses suitable for conversion to residential care homes. Indeed, residential homes are subject to similar fire and safety regulations as small hotels and guesthouses so, in many cases, changes in building use were not time-consuming or expensive (Phillips and Vincent, 1986a; 1988; Andrews and Phillips, 2000; 2002). As a result, many residential-home owners are former hoteliers, while others are nurses who have bought hotels and changed the building use (Andrews and Kendall, 2000).

Nevertheless, as Andrews and Phillips (2000; 2002) note, the 'ideal' funding situation for business growth was to change considerably following national funding changes in the 1990s. The 1990 National Health Service and Care in the Community Act (implemented in 1993) withdrew guaranteed state support for residential-home clients. The broad policy and care aim was to maintain older people in their own homes as long as possible and keep residential homes as a 'last resort' option (Department of Health, 1989; Raynes, 1998). Residential homes have since had to compete among each other for a much smaller number of clients funded by local authority (county council) budgets. A question in much research in the mid-1990s, therefore, was why had the government radically changed its funding policy towards the residential sector?

As Andrews and Phillips (2000) note, studies have commented that the full implications of the guaranteed funding arrangements were not recognised by the government during the early 1980s. Indeed, the numbers of private residential homes in the UK rose from 2255 in 1979 to 7240 in 1986, an annual increase of over 18% (Phillips *et al.*, 1988). Meanwhile, finance provided by the social security budget for residential care for older people, and people with physical and mental disabilities, rose from £6 million in 1978 to £1.3 billion in 1991 (Walker, 1993). Undoubtedly, by the late 1980s private residential homes had become a significant burden on, and element of, the public care budget

(Andrews and Phillips, 2000; 2002). Having residential homes as a 'last resort' option for older people is clearly in line with the broad philosophy of UK care-in-the-community policy and the emphasis accorded to home and home care. However, as Wistow (1995a; 1995b) suggested, an important motivation for the 1990 reforms was the need to control state spending on residential care (Andrews and Phillips, 2000; 2002).

Other research has suggested that tensions existed between private residential care and the government's economic philosophy and vision in the 1980s (Andrews and Phillips, 2002). Certainly, during the 1980s, as part of what was termed the 'enterprise culture' project, many new-right politicians held that private initiative and entrepreneurship were central to Britain's economic revival. Small businesses, in particular, were regarded as being the cornerstone of a healthy and restructured economy. Private residential homes fitted this economic vision, in that proprietors seized a market opportunity presented to them and provided a needed service (Andrews and Phillips, 2000; 2002). Furthermore, the private provision of residential care fitted well with general government encouragement of private provision in a range of health and social-care sectors. However, expansion in the private residential sector was based on weak financial foundations. The irony of a private sector underpinned by, and reliant on, public money meant that it was not a true form of self-generating entrepreneurship (Andrews and Phillips, 2000; 2002).

Furthermore, as Andrews and Phillips (2000; 2002) suggested, another element of Thatcherite ideology in the 1980s opposed the institutional element of residential care. This was a 'new moral authoritarianism' which promoted the positive aspects of family, community and 'traditional' social values (part of the political rhetoric surrounding 'Victorian values' and later 'back to basics'). In this context, providing a financial structure that acts as an incentive for the older and most dependent members of society to be placed in state-supported communal living environments does not sit well with such values (Andrews and Phillips, 2000; 2002). Gradually, during the 1980s, some consensus did emerge between policy-makers and academics that the family model of care was most appropriate, and forms of care which were perceived to lie outside this approach were increasingly opposed (Andrews and Phillips, 2000; 2002).

Whatever the exact motivations behind successive funding scenarios and policy orientations towards residential care, a variety of research has considered residential homes in the current policy climate. Certain studies have used a squarely neo-classical economic approach. Forder (1997), for example, focused on purchaser–provider contracts and the potential for them to create incentives for providers to misrepresent older people's care needs and characteristics in order to increase prices and profit. He also highlighted the possibility for adverse selection whereby providers may exclude high-dependency (higher-cost) clients from their homes. Forder and Netten (2000) followed this line of enquiry by considering the type of contract and its impact upon price. The authors found that residents' dependency characteristics were significantly associated with price, that different types of contract had different prices and risk factors for both purchasers and providers and that block contracts (negotiated with larger providers) encouraged discounted prices. Other studies have shown the new funding environment to have caused business failures, financial hardships and reactive and radical management strategies on the part of private-sector providers (Andrews and Kendal, 2000; Andrews and Phillips, 2000; 2002). Although, in terms of capacity, occupancy and profitability, there are geographical variations (Andrews and Phillips, 2002), in March 1999 overall occupancy rates stood at 87.1% in the residential sector (Holden, 2002). The year 1999, however, witnessed the third annual decrease in capacity across the sector and the sixth year where capacity growth fell short of what would be expected from demographic pressure (Holden, 2002).

More recently, Clarkson *et al.* (2003) considered the latest government proposals to change the funding of older people in residential care. Not as radical as the 1993 reforms, yet still substantial, the proposals are for the final element of central government funding, the 'residential allowance' (a top-up supplementary amount in addition to local authority finance), to be distributed also by local authorities. Hence the proposal aims to remove a remaining financial incentive for residential care and to further encourage home and community care. After talking to care managers, the authors speculated that these price signals and incentives might put added financial pressures on residential homes. However, at the same time, the authors also suggested that client needs, and the sheer numbers of older people unable to be supported at home, mean that the impacts may not be as negative for the residential sector as may be expected.

The role of older people in the future financing of their residential care has received little attention. Hancock (1998) considered wealth through property in old age and found that although housing wealth could be converted in various ways to provide supplementary income, in few cases would it be sufficient to pay for long periods of residential care. Undoubtedly, given the increasing care and residential needs of an ageing population, and public and political debates over the financial responsibility for care, more studies should consider the viability and broad appeal of different funding scenarios. In summary, in 1995 the UK residential care sector provided care for over 170,000 people (Laing, 1995). The future size and structure of national residential care provision is uncertain and much will depend on the extent to which care-in-the-community policies generate viable alternatives to residential care (Andrews and Phillips, 2002). Nevertheless, given even the most optimistic prediction on the success of care-in-the-community, research has estimated that the residential sector will continue to expand slowly (Day *et al.*, 1997). Moreover, in contrast to predictions of a declining private care sector, Laing (1995) predicted that, in the long term, by 2040 the sector will need to double in size to meet the demands of an ageing population (Andrews and Phillips, 2002).

Distributive features in space

As Smith and Ford (1998) commented, research monitoring spatial shifts in the provision of residential homes has been important because geographical concentrations of such services create significant demands on other social and medical services. They also result in variations in the availability of residential beds and thus differential opportunities for older people to enter homes (Smith and Ford, 1998). At the national scale, studies have considered growth and the distributive features of the public, private and voluntary sectors for residential care, often in combination. Early research identified spatial variations in public provision (greater levels of public provision in the north of England) and attributed these patterns to historical influences and the varying policies of local governments (Bebbington and Davis, 1982). Later research expanded these perspectives and considered the national volumes and distributions of both public and independently run residential homes over time. Larder *et al.* (1982), for example, confirmed a greater proportion of public-sector provision in the north and midlands of England and, at the same time, found levels of independent provision to be higher in the south and south-west of England. Overall, however, when taking into account all forms of provision, the south of the country emerged as having higher levels of residential homes. More recently, studies have shown the public sector to have had stopped growing altogether, in some localities to have declined, and a spatial equalisation of the independent sector with greater growth to have occurred in the north of England where, as suggested, it was previously more scarce (Hamnett and Mullings, 1992). Yet more recent research has confirmed a continuation of these trends, namely a more uniform spatial distribution of private provision (Smith and Ford, 1998). However, the significant difference in recent years is that the total residential sector (public + private + voluntary) is not growing as significantly as in the past. Therefore changing spatial trends are largely as a result of changes in proportions of private, voluntary and public provision within regions (in effect, replacement and displacement between the three forms) rather than as a result of rapid expansions in parts of the sector. Nationally focused research has also attempted to distinguish the demographic, social and economic factors that impact upon spatial variations in residential home provision, finding concentrations of over-65s in local populations, and affluence of local populations to be reliable predictors of levels of both public and private residential homes (Larder *et al.*, 1986; Smith and Ford, 1998). Meanwhile, on a related note, Harrop and Grundy (1991) found high densities of older populations to be an indicator of the likelihood of entering residential and other long-term care.

Certain research has focused on spatial features of residential and nursing home care at a smaller scale within particular localities. Pinch (1980) is perhaps the earliest example to highlight considerable spatial variations in publicly provided residential care in Greater London. Subsequent locality studies have demonstrated the complex interplay of variables which impact on the development of residential care in any locality. For example, Ford and

Smith (1995) considered changes in residential care in the south-east of England, between 1987 and 1990. The authors identified continued rapid growth in all sectors of residential care combined but, within this trend, decreases in public-sector provision and increases in independent-sector provision. They also identified saturation of provision in coastal regions and more rapid growth in inland areas where the populations of local towns were ageing most rapidly.

Perhaps the best-known local geographical studies of private residential provision were conducted almost 20 years ago by Phillips and Vincent in Devon, a county with an extremely high concentration of homes. As has been introduced already, these papers focused on local spatial distributions, the impacts of national funding changes, local planning, ownership characteristics, management decisions and choice in care and home life (Phillips and Vincent, 1986a; 1986b; 1988; Phillips *et al.*, 1987a; 1987b; 1988; Vincent *et al.*, 1988). This local research was updated ten years later by a series of papers which focused on the impacts of policy change on local spatial concentrations, business viability, management decisions and attitudes to local policy (Andrews and Kendall, 2000; Andrews and Phillips, 2000; 2002). In both studies, a depth of focus was gained into business features arguably only possible given the small study area and the qualitative methods used. They also provided an insight into local cultures of service provision, an area often neglected by geographical research on ageing. Moreover, local geographical studies have added a depth of focus to, and explore the intimate features sometimes overlooked by, national studies.

Regulation, inspection and quality standards

In the past, the British government has published guidelines for care standards in homes to be implemented by local authorities. Under the 1984 Registered Care Homes Act, the document *Homelife* (1984) was the first code of practice which concentrated on such diverse features as residents' rights, privacy and financial affairs, home administration, physical features and staffing. This document was superseded by *Homes Are for Living in* (1989), which developed a more focused model for the evaluation of care on which local authority registration and inspection could be based and suggested mechanisms and procedures by which local authorities could enact these principles (inspection units were also to be distinct and 'at arms length' from purchasing units). The six stated and basic values of care were listed as privacy, dignity, independence, choice, rights and privacy. Most recently, the 1984 Registered Care Homes Act was replaced by the 2000 Care Standards Act and, under this, *Homes Are for Living in* was superseded by the document *Care Homes for Older People: National minimum standards* (2001). As its title suggests, this document outlined strict and uniform national standards for residential homes. In particular it focused on choice in care, health and personal care, daily life and social activities, complaints procedures and protection, environmental standards, staffing requirements and standards, and management and administration standards. Under the 2000 act, responsibility for regulation and standards was passed from local authorities to eight independent regional commissions.

Not surprisingly, research on regulation, quality and standards has followed legislation and the publication of the above documents and has traced their particular impacts. As Challis and Hughes (2003) noted, in general, since the 1980s, initiatives and documents relating to quality of care have moved from a predominant focus on the structure of homes (for example, allocation of space and fire regulations), to a predominant focus on aspects of the care process, which are now more fully become embedded in quality assessment. A substantial proportion of the research literature on regulation, however, has been rather critical. For example, in the *Homes Are for Living in* era (1989–2000), Phillips and Bartlett (1995) highlighted that the private sector often called for a fairer system and expressed concerns over local authorities being both inspectors and purchasers. The Social Services Inspectorate (1994), meanwhile, highlighted differences in the performance of local authorities in terms of numbers of inspectors, the quality of self-monitoring, and volume and quality of information provided to home owners. Soon after, a Council and Care report by Goward (1995) expressed even more severe criticisms, citing unnecessary bureaucracy in some local authority inspection units and inconsistencies between inspection units in terms of required standards and the quality and regularity of inspections. In the same year, the 1995 Wing Report investigated how registration and inspection is perceived by residents and their relatives. It concluded that residents and relatives were often ill-informed or unaware of inspection units and their role and called for improved complaints procedures. Testing the quality of care and quality of life in homes, Challinger *et al.* (1996) found significant differences to exist in both and called into question the effectiveness of the regulatory systems of the time.

At the time of writing, because the current regulatory era has only just begun, there has been little time to trace the impacts of national minimum standards, in terms of meeting their objectives and more generally on the residential sector and residents. However, Darton *et al.* (2003) discussed initial evidence that some homes have recently had to close due to their inability to reduce costs while meeting new standards. Furthermore, Holden (2002) has noted that the new national standards have produced controversy. For example, the author suggests that the directive that residents should be given a choice of single room (of minimum area $10m^2$), and that only 20% of any home's total rooms may be double rooms (of area $12m^2$), means that many private homes will have to reduce capacity and may face financial difficulties and that many local authority homes do not meet these standards and may have to close. Holden speculated that if standards encourage a new wave of building, because of size requirements, it is likely that more institutional larger homes will predominate.

Although the above studies on registration, inspection and quality standards are undoubtedly important, they either take a perspective concerned with the financial health of the residential sector or consider the experiences, needs and welfare of older people on their behalf. An alternative perspective, however, is taken by Raynes (1998), who, by talking to older people themselves, considered their potential role in quality specification and their opinions regarding the contract for the local authority purchase of their residential care. According to the older people Raynes interviewed, in order of frequency of responses, what makes a good home is social activities in the home, social activities outside the home, choice of good food and drink, kind staff, access to one's bedroom, pleasant company and friendship of other residents, continuity of staff, physical comfort of surroundings, access to support services, personal safety, aid for self-care and overall size of home. Furthermore, Raynes identified some other factors which enabled the views of users to be incorporated into service specification. They included time and energy, attention to detail, risk-taking, sense of humour, food, pleasant venues, transport, access to a phone, a steering group, continuous communication, focus on outcomes and partnership between equals. In the spirit of giving older people a voice in gerontological research, this type of approach is advocated both by Raynes and here as an important component of both evaluating the impact of the current national standards and setting future standards on residential care. Indeed, obtaining older people's own views on the places where they live should arguably be a priority in both residential-home research and formal quality assessment.

Residents

The movement of older people in and out of residential care is an important consideration of research. Netten *et al.* (2001), for example, considered the appropriateness of placement decisions and in particular demonstrated differences in proportions of nursing and residential placements between local authority areas. The need for residential care has also been viewed in terms of dependency and wider push-and-pull choice factors (Reed *et al.*, 1998). Reed *et al.* (2003), for example, summarised four types of typical relocation, which are not necessarily mutually exclusive: preference relocations (where older people are in complete agreement), strategic relocations (planned by older people to pre-empt dramatic changes), reluctant relocations (with which older people disagree, but which changing needs dictate) and passive relocations (in contrast to the former, older people accepted or did not question their move). Furthermore, under circumstances where a move into residential care is necessary, Eales *et al.* (2003) found that older people particularly valued the physical setting of the home and its location with respect to local community and neighbourhood and valued also a choice available between homes.

The dedicated study of residents goes beyond locational considerations and has been one of the longest-standing traditions in residential-homes research. The theories and foci used have changed substantially. As Bochel (1988) suggested, during the 1970s and 1980s three main approaches were taken in the study of residents. The first approach considered broad characteristics of groups of residents such as their age, gender, funding sources and admission routes (Tibbenham, 1985). The second approach considered residents in relation to their environment. A substantial proportion of the literature was focused on residents' quality of life, often in relation to home administration (Davis and Knapp, 1981; Peace *et al.*, 1979). The third considered design features and their impact on residents. As Bochel (1988) commented, a difference of opinion emerged in this early literature. Willcocks *et al.* (1982) found that design did impact on residents' activities and interactions while Lipman and Slater (1976) and Lipman *et al.* (1979) found no relationship to exist between building design and residents' behaviour. More recent research has confirmed this relationship, supporting the view that design influences both residents' well-being and the quality of care by staff (Keen, 1989; Netten, 1993; Calkins, 2001). As Barnes (2002) commented, micro perspectives have focused on one or only a few design characteristics while macro perspectives have focused on the setting as a single unit and consider the impact of the whole setting on individuals. In particular, wayfinding research (how people negotiate their immediate environments) has been applied to residential care settings (Passini *et al.*, 1998; 2000). As Barnes (2002) suggested, other research considers the importance of gardens for residents' privacy, activity and stimulation (Brawley, 2001); the importance of artificial and natural light (Okumoto *et al.*, 1998; Kolanowski, 1990) and design features and residents' privacy, focusing particularly on notions of personal space, territory and intrusion (Duffy *et al.*, 1996, Keen, 1989). Willcocks *et al.* (1987), for example, pointed to how 'the concealment of frailty' attained in one's home is not possible in the communal areas in residential homes.

Despite recent interest in the impact of design features, to date, there has been a shortage of appropriate, comprehensive and objective instruments for assessing the outcomes of building design (Barnes, 2002), As Barnes (2002) suggested, one known assessment that could be applied is the multiphasic environmental assessment procedure (see Moos and Lemke, 1996), which focuses on residents and staff characteristics, physical features of the home, the social climate, accessibility within the home and to outside services, overall facilities and safety. However, as Barnes argued, this tool is most appropriate when applied to larger settings. Another assessment tool with specific reference to dementia care settings is the professional environmental assessment protocol (see Lawton *et al.*, 2000), which is based on attributes of place experience (Barnes, 2002). However, again this tool is focused on a particular client group that constitutes only a proportion of residents in homes. Furthermore, it is not specifically designed for application to residential homes (Barnes, 2002).

Over time, the approach to residents has become more expansive than foci on design features. Arguably, it is one of the most progressive areas of research on residential care, methodologically becoming concerned with giving residents a voice in research and making their views on residential living heard. Research here has considered the transitional period once an older person enters residential care. Reed and Roskell-Payton (1996), for example, considered ways of adapting to life in nursing and residential homes. The process of adapting was found to involve extensive social activity, negotiating complex social conventions, constructing familiarity and managing the self. Further to this, both conventional wisdom, and some research evidence, has suggested that older people resident in homes are only there because they have to be and would like to return home if this were possible. However, Noro and Aro (1997) instead found that, when asked, most residents would prefer to stay in residential care, where they felt secure and well cared for.

Residents' social interaction in residential homes has been a limited area of research. Hubbard *et al.* (2003) commented that low levels of interaction and social activity have been found in residential care settings (McKee *et al.*, 1999), while Fox (1999) considered control, resistance and timing in residential care and found that time in residential homes is something which has to be filled. However, Hubbard *et al.* (2003) caution against perceiving all interaction as good interaction because when interaction does occur and it can equally be hostile and unfriendly as reciprocal and caring (Powers, 1996, Reed and Roskell-Payton, 1997). Social interaction has also emerged as an important feature in debates about the institutional nature of residential care. Reed *et al.* (2003), for example, suggested that research has begun to challenge the perceived passivity in being a resident, and has begun to convey how residents shape and make their own lives within homes by knowing the environment, talking to and learning from other residents, and engaging in social activities (Reed and Roskellin Payton, 1996; Oldman and Quilgars, 1999; Raynes, 1998). This and other research form part of the 'residential versus community care debate', discussed in detail elsewhere (Andrews and Phillips, 2000; 2002) and in the second chapter of this book. Hubbard *et al.* (2003), however, commented that despite increasing research on home life, there is still very little understanding in research of residents' jokes, irony, sources of hostility and sex and sexuality (Bauer, 1999), the dominant view with regard to sexuality being that residents are without sexual identity, interests, needs or capabilities and that any such expressions should be repressed (Hubbard *et al.*, 2003; Brown, 1989).

Residential homes as places of loss and bereavement require more research attention. To date, little has been undertaken but research in other living contexts indicates its potential and importance. Hockey *et al.* (2001), for example, focused on older people living alone in their own homes or sheltered accommodation, and on grief, loss and changes in use of domestic space. For these authors, bereavement was a spatialised experience, older people's own homes being 'store-houses' of memory created through years of routinised movements in domestic space. Such research raises questions about residential homes and the meanings and attachments that older people may attribute to them in life and in death.

Private ownership and the management of care

Many studies of residential care have explored features of private ownership. Early research considered basic business features including length and forms of ownership. Among other things, husband-and-wife partnerships tended to be a common ownership and management scenario (Phillips and Vincent, 1986a; 1986b; 1988; Phillips *et al.*, 1987a; 1987b; 1988; Vincent *et al.*, 1988). Hence the term 'family business' has often been applied to residential care, not only because of this but because the notion of family has been found to also exist as a caring philosophy and as a marketing tool (Phillips *et al.*, 1988; Andrews, 1997). Divisions of labour have also been identified within 'family work teams', including women or wives undertaking more 'hands-on' caring tasks, while men or husbands undertake more maintenance and managerial duties. Children and other relatives have also been observed to undertake significant amounts of unpaid labour (Phillips and Vincent, 1986b; Phillips *et al.*, 1988). The previous employment of proprietors has also been researched, and their pathways to residential home ownership. As suggested, studies have found a substantial proportion to be property developers, hoteliers and nurses (Phillips *et al.*, 1988; Andrews and Kendall, 2000). Indeed, theoretically the ownership of private residential care has also been characterised as being 'petit bourgeois' in nature, involving modest amounts of property ownership, high levels of financial insecurity, low levels of direct employment generation and business owners undertaking the hands-on work themselves (Phillips and Vincent, 1986b; Andrews, 1997). Research has also considered the experience of ownership, finding high levels of stress, burnout and turnover (Andrews and Phillips, 2000). Management decisions, plans and aspirations have been a substantive area of research; studies in these areas, for example, have found advertising and marketing to be undertaken at a very modest level. Other research has shown that under policy and funding changes, and in the face of concurrent financial difficulties, proprietors have often implemented quite radical and reactive management strategies including the loosening of admissions and discharge criteria, business diversification and reductions in spending (Andrews and Phillips, 2000; 2002).

Beyond Phillips' and Vincent's early observations, other research has more squarely investigated notions of home, family and privacy in residential care practice. Peace and Holland (2001), for example, questioned 'homely residential care' and whether it is a contradiction in terms. The authors suggested that residential homes always have to balance the needs of individuals and groups and empirically studied small homes (less than four beds) to see if residential homes can replicate some experiences of independent living. The authors found there to be a certain 'homeliness' in these small homes, more than any other type and scale of residential care but, even on this scale, tensions were still found to exist between home and institution. Small homes were found to be capable of producing the best and worst of home and institutional worlds. Bland (1999), on the other hand, considered independence, privacy and risk and contrasted two different approaches to residential care, a private home run by former hoteliers (a service approach) and a local authority home (a social care approach). The author provided some tentative conclusions as to why a service rather than social care approach may be better at preserving residents' autonomy and privacy. She commented that in the private sector residents are treated more like customers. In contrast, the care approach in local authority homes reflects both residents' and providers' statutory duties.

Most recently, Holden (2002) considered concentration (i.e. the role of big business) in private long-term care but commented that it is more often found in the nursing home sector (as opposed to the residential sector), because economies of scale can be better applied to these businesses. Furthermore, the author commented that financial barriers to market entry have risen for all private long-term care providers in the UK, not only because of changing regulatory criteria, but because banks are less willing to lend money to small businesses under the current market conditions. The author concluded by expressing concerns over the effects of increased ownership transfers on staff morale, continuity of care, disruption to residents and transfers of residents. Holden commented that if in future large companies do end up dominating the residential care market, taylorist style standardisation of care may lead to more institutional-style care, a reduced choice for consumers (with respect to types of home and care ethos) and a decline in care quality, particularly if localised monopolies occur.

Staff

Many studies focus, in part, on the issues that effect or involve the staff of residential homes. However, only a small volume of research has focused exclusively and in depth on the staff. Studies by Phillips and Vincent in the 1980s, and more recently by Holden, outlined some of the basic features of the labour force, namely that a lot of care staff were young, female, low-paid and non-unionised and displayed a considerable amount of transience between different care homes, or in and out of the sector (Phillips, *et al.*, 1986; 1988; Holden, 2002). In the context of funding changes and concurrent financial hardships (and prior to the introduction of the national minimum wage), the follow-up study by Andrews and Phillips found that almost half of proprietors had not given their staff wage increases as a partial response and coping strategy (Andrews and Phillips, 2000). The above studies have also highlighted at the potential for transient or disenchanted staff to impact negatively upon the home life of residents (Andrews and Phillips, 2000). More recently, Holden (2002) found the introduction of the national minimum wage to be having an impact. On average, care assistants' wages had increased but an uneven national geography of wage levels reflected local wage markets. Furthermore, meeting new wage demands was found to cause an uneven geographical increase in fee levels for residential care. For example, Holden found that, in Yorkshire, homes had to increase fees by more than seven pounds a week to cope with the financial demands of the national minimum wage, while in London homes only had to increase their fees by 80p. Holden also found that European legislation had impacted on the sector. In particular, the 1998 Working Time Directive had increased wage bills due to the new staff entitlement of three weeks' paid annual leave. Other research has focused on stress among staff. Baillon *et al.* (1996), for example, focused on the factors that contributed to stress in publicly-run homes, finding organisational and procedural duties to be more stressful than care-related duties. Supporting these findings, Jenkins and Allen (1998) found burnout and distress among staff to be associated with the quality and quantity of social interaction between themselves and residents. The authors also found perceived involvement in decision-making among staff to be positively associated with the perceived quality and quantity of staff–resident interaction.

The above studies have demonstrated the importance to staff of caring, however, and perhaps surprisingly, very few studies have considered the caring activities of staff and how they may be improved. Moxon *et al.* (2001) considered the feasibility of involving care staff in reducing depression in people in care homes. In their intervention, mental-health training was provided. Subsequently the detection of depressions improved over time as well as depression being reduced in some residents. This is one of only a few studies of practice but it does showcase the importance of research in this area.

Discussion

Although geographers have engaged with, and contributed substantively to, the macro-scale categories of residential-home research outlined above (for example, policy, financing, regulation), their contribution to micro perspectives on homes or place (and residents, proprietors, staff) has been more limited. Indeed, most of the research I have discussed on these last-named categories has been undertaken by social gerontologists with an emphasis on a particular feature of place (for example, an activity), rather than treating place as a complex set of situated social dynamics; a symbolic and social construction. Collectively, their studies have effectively unpacked and provided a rich description of residential homes. However, there is still room for geographers to take a more holistic view of residential places and theorise and engage directly with them in individual studies, effectively joining up the different dimensions of place highlighted by social gerontologists.

As far as the future research agenda is concerned, one important cross-cutting theme is to engage with wider critical debate in social gerontology on images of ageing in both the negative contexts of stereotyping and stigma and the positive contexts and the positive features of ageing (see Bytheway, 1995; Blaikie, 1995; 1997; Featherstone and Wernick, 1995). Negative societal perceptions are thought to originate from a variety of sources including media images. However, the typical social policy, health service and academic focus on the 'problems' of old age is thought to be just as much to blame for the wider perceptual problem (Conway and Hockey, 1998). Recently, McHugh (2003) has argued for a spatial perspective in this field of social gerontology, as he considers notions of ageing to be very much tied to places. McHugh posits that spaces and places are not passive containers of or inert backdrops to social activity but that geography is sometimes invisible, taken for granted, and can be integral to images and perceptions. He argued that even positive place-related images (e.g. retirement areas and communities) might be counter-productive. Extending this perspective, residential homes and their image appear to be a very relevant area of research. However, to date, only a single study has come close, being focused on nursing rather than residential homes. Indeed, Gubrium and Holstein (1999) stated that because the nursing home is a paramount site for the care and custody of old people, it also affects how we think and talk about ageing. To these authors, it is an anchor for embodiment that frames and represents the body in powerful ways that other sites, such as hospitals, do not. For example, the authors state that a hospital's image or 'gaze' does not have such a culturally focused relevance to old age and that a hospital is connected more to recovery, as opposed to continued physical and mental decline. The authors explain that the discursive anchoring of setting and body works like a categorisation device; particular negative body types and social situations are perceived to always exist within the setting. Furthermore, the authors talk about the reinforcement of this image through the de-privatisation of embodiment, because in residential homes physical frailty is displayed in public view. This study certainly demands further research, particularly on the practical causes of the stigmatisation of residential homes, and on potential remedies. Relevant questions include: what images do different societal groups hold? How do different groups in society feel about the image held by residential homes? Within this, how do residents feel, their families, and owners? Certainly, such direction would bring the study of residential homes into line with the emerging critical concerns of social gerontology and would effectively 'emplace' research on stigma and image while at the same time help inform policies and actions to improve the image and operation of residential homes.

More generally, future geographical research may span some of the traditional sub-disciplinary strata of human geography. In terms of the social geography of residential homes, there is need for geographical research on different age groups of older people, older people with different levels and types of dependency and different income groups and their experiences of different types of residential home with different caring philosophies and caring practices. In terms of the culture of residential homes, there is also a need to investigate more substantively the co-production of residential-home culture and the 'making' of place. Many stories remain untold, including home traditions (official and unofficial) and the stories and beliefs of residents and staff. Political geographies remain an important consideration. Building on current national and regional perspectives, research could re-theorise accessibility. It is crucial that this research be conducted within a framework that acknowledges the interaction of the different scales of political places, and the often messy and contingent nature of place as an ongoing process of production from the national governmental level to town planning. Meanwhile, economic geographies could focus not only, as they have done, on the distributive features of provision and the labour market, but also on the social and power relationships embedded within homes as workplaces; in terms of dominant cultures of production, their counter-cultures and the everyday interactions of owners, staff, residents and families through performative labour relations. Another area for research could consider how local forms of residential care may be located in specific versions of urbanicity and rurality and combine with other features to make these places. Feminist geographies of residential care, which to date have been few, could investigate how gender relations and residential homes are mutually constructed. In terms of history, researchers could investigate how residential care is embedded in regional and local histories and local historical identities (particularly coastal towns), how residential homes effectively contribute to the gradual 'becoming' and 'making' of places and how they are contested in the local histories of places. Evidently a wide-ranging research agenda still exists for a well-researched, but ever changing, form of older people's accommodation and care.

6

HOME AS A NEW SITE OF CARE PROVISION AND CONSUMPTION

Janine Wiles

> What needs to happen, is that new models of care, alternative models of care, need to do away with the slippery slope [a direct line from residential to institutional living], and allow people to remain in highly residential settings where the services that they need are being brought to them, in some way, shape or form, and I believe that we haven't begun to tap the potential for creativity and innovation for bringing services to people who are remaining in settings of their choice, and I also firmly believe that creative applications of that can also be done in very cost-effective ways.
>
> Dean Toni Sullivan, Sinclair School of Nursing, University of Missouri-Columbia, addressing Joint Interim Committee on Aging Hearing, July 28, 1998

There are two sides to the literature on the provision and consumption of care at home. Critics argue that increasing pressure on financially constrained, and sometimes qualitatively inadequate, resources leads to severe shortcomings for both service users and providers. Yet there is a generalised sense of support for the idea of care that takes place in a home environment, which is seen as having potential therapeutic benefits for those receiving health care and their social support networks, and the potential for growth. In this chapter, I will explore literature that has critically examined the home as a 'new' site for the provision and consumption of care for the elderly.

Demand for home care is increasing

Demand for home care is increasing everywhere (Hutten and Kerkstra, 1996). Growing proportions of very elderly people and improving life expectancy for those with chronic illness lead to greater demand for care. Home-based care, particularly for seniors, is now a substantial and rapidly growing element of the health care system in many Western industrialised countries (Hutten and Kerkstra, 1996; Williams, 1996; Naylor, 1997). People now live to be much older than was previously the case, but there are also increasing incidences of the diseases and illnesses associated with old age (Moore *et al.*, 1997; Ward-Griffin and Marshall, 2003). Yet, at the same time, more than ever before there are increasing numbers of relatively healthy older people, and it is not always appropriate to imagine care for these people in institutions and nursing homes, whether based purely on a medical model or even on a more advanced holistic psychosocial model.

There have also been changes in the geography and social nature of 'the family'. Changing historical circumstances cause variations in both people's need for family support and the capacity of family to provide it (Finch, 1989; Hareven, 1991a). For example, migration, urbanisation and female participation in the labour force lead to transitions in living arrangements, so that families are now smaller and more geographically dispersed (Martin, 1989; Moore *et al.*, 1997). More elderly people live alone, especially elderly women, and this latter group makes up a great proportion of elderly people who are economically vulnerable (Krivo and Mutchler, 1989; Moore *et al.*, 1997). Moreover, in the late twentieth century women, the 'traditional' care-givers of elderly people, are more likely to have some form of paid employment (Doty *et al.*, 1998).

Rapid improvement in technologies for home care has also had an effect in changing locations for the provision and consumption of care from institution to home (Ward-Griffin and Marshall, 2003). Telemedicine, an umbrella term including such activities as the sharing and maintenance of patient health information, and the provision of clinical care and health education to patients and professionals via telecommunications over a distance, has already radically transformed the nature of home-based care provision (Bauer, 2001). Telemedicine is popular because it is perceived as having the potential to reduce the costs of home care while improving quality and integration of a range of existing services, and enabling the expansion of home care into rural areas (Health Canada, 1998).

The shift towards 'ageing in place' and home-based care is further augmented by expanded definitions and understandings of the nature and meaning of health (WHO, 1946), which incorporate social and emotional well-being as well as physical and medical attributes, and by a re-emphasis on prevention relative to curative medicine.

The provision of care at home is subject to ongoing reform

These changing demographic, social, and technological circumstances are intimately bound up with changes in the delivery of health care services to elderly people. There have been reforms to welfare state provision in many countries, and changes in formal care services such as closure of institutional care for seniors (Ward-Griffin and Marshall, 2003). Waiting lists and shortages of staff are problems everywhere (Hutten and Kerkstra, 1996). Given the context of economic recession and budget cuts, many governments are left facing the practical and political issue of how to most efficiently and effectively provide care to elderly people and make financial savings (Sundström and Tortosa, 1999). For some governments, and many other groups, an equally important issue is how to provide care to elderly people equitably.

In response to these issues of growing demand and rising costs of the provision of home care, and shaped by ideological perceptions of both the nature and scale of problems and the need for solutions, there are general trends emerging in the provision and reform of home care for elderly people. There has been a shift towards support for the desires of elderly people and their families for 'age in place', and the home has re-emerged as a key site for the provision and consumption of care for the elderly.

One key element of this shift in care for the elderly is a general trend towards the decentralisation of responsibility and control over services (Sundström and Tortosa, 1999; Williams, 1996). Responsibility is being shifted from the level of national government to more localised government, from health ministries to social services, and outside of these sectors altogether to become the responsibility of a sometimes *ad hoc* range of not-for-profit, charitable, for-profit, and community-based organisations. This decentralisation means that even within a single country there may be wide variations in the range and amount of services provided, or per capita funding for home care (Health Canada, 1999). Because the mandate for home-based care of the elderly often falls somewhere between health and social services, the fact that in many jurisdictions these are separate entities has proved to be a major problem (Hutten and Kerkstra, 1996). Moreover, some argue that because the funding of home care is less visible than that of institutions, home care services can more easily be cut back gradually and less visibly and with less likelihood of public attention or resistance from organised groups (Sundström and Tortosa, 1999).

Within the direct, publicly funded provision of home-based support, there is a trend towards adopting some form of 'single-window' or 'one-stop-shop' access point, where the needs of potential users of home-based care can be assessed and they can be directed to the most appropriate mix and intensity of support services. Single-window assessment is frequently combined with ongoing management, often on a case-by-case basis. This is convenient for users, many of whom are unaware of the range of support that can potentially be accessed from, and used at, home, and useful for service providers who are, in theory, better enabled to provide a more integrated and coherent suite of support services and to monitor the effectiveness of those services.

Crucially, however, such structures for service allocation provide a critical point at which funders can invoke the rationing of home-based support. This generally takes place through the imposition of one or a combination of strategies such as tightly capped budgets, strict needs assessment and eligibility criteria, service user fees, ongoing reassessment and management, or direction towards other non-publicly funded services both for-profit and not-for-profit, formal and informal. Eligibility criteria are often related to the potential availability of a family member to provide care, or may be connected to ability to pay privately for services. Capped budgets mean that publicly funded services are often not provided to all those in need, but limited to those defined as most in need, whether because they are most frail, most financially vulnerable, or socially vulnerable and without family available to provide care. That is, it could be argued that in spite of often well-intentioned ideas, the primary motivation in implementing single-window allocation frameworks in practice has been to control spending and respond to financial crises, rather than the equitable distribution of resources or attempts to prevent health crises and promote the health and well-being of those at home. Benefits to 'consumers' and providers have become side effects, or have disappeared altogether, secondary in importance to the need to control spending.

Care is provided by a range of groups

There are at least two broad kinds of care provision received at home: formal (usually by workers who are paid to provide care) and informal (unpaid assistance, usually meaning care by family, or friends and neighbours). While many researchers have treated these as two separate entities, more recent work is critical of this dichotomy. Dividing care provision into formal or informal is seen as too simplistic an approach, especially when informal care is treated as if it were unproblematic and not 'work' (Ward-Griffin and Marshall, 2003), or when the expertise developed by 'informal' or lay care-givers through caring for an individual is not recognised or is devalued (Pickard *et al.*, 2003).

This significant critique notwithstanding, most support for frail elderly people at home is provided on an informal or lay basis (Larragy, 1993), and a key feature of the provision and consumption of care at home in general is that most is provided on an unpaid basis. Estimates suggest that between 75 and 90% of care received by seniors in the community is provided informally by family and friends (Abrams and Bulmer, 1985; Aronson, 1990; Health Canada, 1997). Generally responsibility tends to fall to one primary care-giver, less frequently with support from broader informal networks (Bond *et al.*, 1999; Chappell, 1991). Much of this care is provided by elderly people themselves, debunking myths about the general dependence of elderly people and the subsequent burden thus placed on ageing societies. For example, a recent report commissioned by Health Canada suggests that 25% of those providing care to a chronically ill or frail family member at home on a long-term basis are at least 65 years old, while 57% of those receiving care were aged 65 years or older (Health Canada, 2001a). Other research suggests that only a minority of people actually receive formal support (Stone *et al.*, 1987). The bulk of care is also provided by women, to women; the provision and consumption of care at home is a gendered phenomenon.

Publicly funded services themselves may be provided by individuals employed directly by employees of government agencies, or by employees of other institutions, such as for-profit private organisations or not-for-profit organisations such as charitable agencies and institutions. Sometimes single-window agencies draw on a mix of private and government-funded service providers. Home care provision in Scotland, for example, has evolved into a quasi-market system, incorporating a mix of formal and informal care, some of which is provided by local government, and some through private agencies competing for contracts from government (Curtice and Fraser, 2000).

The for-profit provision of home-based care for the elderly, and indeed in general, is one of the fastest-growing markets in health care. Market-based approaches to care such as those employed in the United States suggest there is much potential for growth and investment as populations age and clinical protocols allow for a greater range of services and therapies to be provided at home. Yet as an industry home care is highly fragmented, and private home care providers tend to be small and locally based. Even in the relatively well-developed market system of the United States, for example, home health care (HHC) agencies tend to be small and not highly capitalised (Scully *et al.*, 2003). Although home care avoids some of the large capital costs associated with institutional care, and costs are substantially labour-oriented, shortages of skilled professionals place pressure on agencies and the small size of many agencies makes it difficult to attract investment. In the United States the Balanced Budget Act of 1997 severely impacted on the previously rapid growth of the home care industry (i.e. from $3.3 billion of Medicare expenditure in 1990 to $18 billion in 1997) and led to the closure or merger of over 3000 agencies. The introduction of the Medicare Home Health Prospective Payment System in 2000 stabilised the industry and has been linked to current growth rates of between 5 and 10% annually, along with demographic trends and relative cost advantages (Scully *et al.*, 2003). The American experience suggests that government policies and involvement in subsidising home care are thus crucial to creating and maintaining stability, even in a 'market' system of home care.

As a new place for the consumption and provision of care, home needs to be critically understood

The relationship between people and their environments is a central element of the ageing process, and the need for a critical engagement with the social and spatial context of ageing is further emphasised by political and public interests in the idea of 'ageing in place' (Binstock, 2000). Although there is not much research on the nature of the home as a site for the provision and consumption of care, researchers are pointing to the need for research that would enable a better understanding of the meaning and experience of home as a site for care in the context of restructuring the health care system (McDaniel, 2000; Williams, 2002). Homes are not neutral containers in which the provision and consumption of care can take place. Rather, they are already infused with rich meanings, histories and significance, which the provision and consumption of care disrupts, displaces, recreates and reproduces (Wiles, 2003a).

Home is not necessarily a new site of health care provision and consumption. In an era of de-institutionalisation there is a tendency to hark back to an idealised memory of a 'golden age' of family and home care as a way to encourage or make sense of care at home (Hareven, 1991b). Yet feminist research challenges these nostalgic assumptions about historical patterns of intergenerational support, suggesting they may not be so strong as is widely believed and indeed are much more complex and patterned by relationships of power and resistance, and shaped by gender and age relations (Abel, 1992; McDaniel and Lewis, 1997).

Certainly in the past doctors would typically visit people at home, and much care was also provided by family members and those associated with a household. Historically, religious and charitable orders would also visit people in their homes. As early as 1897, for example, the Victorian Order of Nurses was inaugurated in Canada to be an order of visiting nurses in remote areas and rapidly growing towns and cities to care for sick people in their homes. In the context of publicly funded services struggling to match demand to their budgets, charitable organisations such as this are once again playing an important role.

Feminist theorists often approach the idea of home from the need to critically re-theorise oppressive dichotomies between public and private, formal and informal, paid and unpaid work (Angus, 1994). Much of this work challenges notions of the home as a private space where primarily reproductive activities take place, and instead examines the significance of home spaces in shaping and confirming gender and sexual identities as somehow 'natural' (Bowlby *et al.*, 1997; Domosh, 1998). The supposed 'public–private' divide and the 'appropriate' activities that are associated with it create and maintain the negotiation of gendered roles and power relationships, not only between public and private spheres, but within the 'private' space of the home. For example, caring is seen as most appropriately taking place in the 'private' domain of the household, the private space also identified as 'female'; thus caring work is perceived as private, female work.

A range of feminist research points out that the home, rather than being a sanctuary and a haven, is actually a site of oppression for many women (Rose, 1993). Home is the site where capital exploits the unpaid reproductive labour and the caring of women in 'traditional' stereotyped gendered roles, and where the welfare state makes and maintains distinctions along the lines of familial gender roles, further trapping women and men.

Moving care into the home and community thus has significant consequences. The experience is often complex and contradictory, for the elderly people who are the focus of care, and for the informal and formal providers of care. Milligan notes the blurring of public and private space within the homes of people providing care. She argues that domestic space has merged with public services to become a site of caring and often the sole locus of formal and informal care provisioning, suggesting that community-based care 'may be more aptly described as an institutionalisation of . . . private space' (Milligan, 2000: 55). For example, Twigg points out that the temporal rhythms of service providers and receivers are different, and that this is exacerbated in the home:

> Service delivery operates according to temporal structures that have their roots in the rationality of economic production. It is ruled by clock time and the need to construct efficient rosters. But domestic time is constructed differently. It has its roots in social meanings and the needs of the body: getting up and dressed, eating meals, going to the bathroom. They provide its temporal rhythm, and its nature is different from that of the clock time of work schedules. As a result there is often a conflict between the demands of efficient service provision and the needs and wishes of older people.
>
> (Twigg, 2000b: 390)

Her work provides a good example of the home as a place full of meaning and emotion to those who live in it, compared with the home as a space which is convenient (or not) and cheaper for those providing formal care. A Canadian-based project extends this idea to the spatiality of care, examining the consequences for care recipients, households, homes, and workers as homes become primary settings for formal health care and social support service provision (e.g. England, 2000; McKeever, 2001a; 2001b). They emphasise the relationship between the material aspects of home and social power dynamics as homes function simultaneously as personal and family dwellings and sites for complex, labour-intensive health care work. For example, the needs of visiting nurses and family members may conflict, as the former seek a hygienic and convenient work space while the latter may emphasise aesthetics and the pragmatic needs of home over a sterile nursing environment, or have difficulty dealing with the daily intrusion of health workers and their needs.

Twigg also looks at the spatial ordering of care at home by unpacking the idea of privacy at different scales (between home and outside home, within home, and around the body) and exploring how this relates to power in the context of care at home (e.g. through class, gender and place; Twigg, 2000a). She explores the meaning of home, and the implications of provision of care in the 'context of the structured intimacy of home' (Twigg, 1999). She concludes that home is

> about privacy, security and identity. It embodies the self, both in the sense that it is the concrete extension of the self and in that it contains and shelters the self in its ultimate form of the body. Formal care services in entering this territory need to negotiate their way through these structures, transgressing boundaries and reordering social categories, but in ways that recognise the power that lies within them.
>
> (Twigg, 1999: 10–11)

Twigg's work is designed to understand connections between the body and community care, to expose the embodied and physical nature and meaning of the provision and consumption of care at home. She helps us to reflect on the dissonance between the space of the home and the disembodied, abstract ideal 'community' space of the service and policy sector.

We cannot understand homes either as simply material locations, such as houses or apartments, or as simply social and emotional ideas. A growing body of research points to the connection between the material adequacy of homes as places to provide and consume care, and the deep social and emotional implications of the home becoming a location for the provision and consumption of care (Mowl *et al.*, 2000; Wiles, 2003a). Care at home is often related to significant social isolation as the nature and use of the home changes and the ability of care recipients to travel, and for others to visit, are compromised (Bamford *et al.*, 1998; Wiles, 2003a). Consumption and provision of care at home may also be related to necessity (and ability) to make modifications to the physical environment of the home, which in turn shapes the experience of care at home (Messecar, 2000).

Homes as material objects and symbolic entities are shaped and reshaped by their occupants over time in response to changes in the individual's life course and the social context within which they are set, both immediate urban or rural contexts, and general social and welfare reforms (Perkins and Thorns, 1999). The location of homes has implications for care, whether access to formal support services for homes located in isolated rural areas (Morris *et al.*, 1999), or location relative to resources such as day-care centres or the accessibility of local transport (Milligan, 2000).

Care at home does not take place in a neutral space, nor are all homes alike in the ways that care is able to be provided and consumed and how it is experienced. Homes are at once physical, symbolic, social and political, and always connected to other spaces. This may be directly (e.g. the physical location of a home) or in terms of meaning.

Consumption and provision of care at home is shaped by complex emotions and necessities

People feel strong attachments to homes as places, not only to the physical setting itself but to the things, experiences, memories and expectations embodied therein (Marcus Cooper, 1995; Rowles, 1983a; 1993; Rubinstein, 1989). Being at home is strongly associated with a sense of autonomy and independence, of having control over one's daily activities and events, one's body, individuality and social status (Golant, 1984: 175). Kontos argues that home thus 'affords independence by defining a space that is controlled by and is uniquely the domain of the individual', so that home as a place is integral to how old age is experienced and constructed (Kontos, 1998). Yet although home as a place is often imagined in such ideal terms, the lived reality of home rarely matches up to this ideal (Bordo *et al.*, 1998; Perkins and Thorns, 1999).

Perhaps just as crucial as the reasons for living at home are negative perceptions of the other options available to elderly people, notably institutional spaces such as a retirement, nursing or long-term care home (e.g. Cloutier-

Fisher and Joseph, 2000; Earhart and Weber, 1992; Greenwood, 2002; Ryke *et al.*, 2003). Families also express negative responses to the idea of institutions (Courts *et al.*, 2001; Health Canada, 2002). There is concern about the nature of institutions (whether as social spaces, or because of the risk imposed by concentrations of disease and bacteria in hospitals). The symbolic imagery of the institution is as strong and evocative as that of 'home'. For many, the difference is a matter of 'living' at home as opposed to 'waiting to die' in an institution. Concern about institutions is balanced against beliefs about the positive relationship of the home place to health (Ruddick, 1994). As imagined spaces, institutions conjure up ideas of lack of autonomy, independence and privacy, even if the lived reality of those spaces is often very different and many institutions have moved away from the medical model to develop a social model of care incorporating a more 'home-like' environment.

Policy-makers and service providers respond to these concerns and government rhetoric tends to construct home as the preferred locus of care (Anderson, 2001; Health Canada, 1997; 2002; Joint Interim Committee on Aging, 1998). From the perspective of providers and policy-makers, reasons for providing services to elderly people at home also include a desire to improve standards of care and quality of life and to maintain people in their homes for as long as possible. Provision of services at home is also motivated by shortages or decline of resources in public institutions such as nursing homes. More broadly, providing services at home is perceived as being more cost-efficient and more humane than providing services in institutions, particularly in health care systems facing increasing costs and demand (Sundström and Tortosa, 1999; Teeland, 1998).

Relationships between formal and informal carers take on new meanings when care is provided and consumed at home

The provision of formal support itself may be motivated by the goal of relieving family care-givers, or to benefit ill, frail or disabled seniors in the community, or to prevent institutionalisation, or a mix of all three (e.g. Donaghy, 1999; Zarit *et al.*, 1999). Formal support may be provided to complement informal support provided by family, or on the basis that families are not expected to provide much care, or on the understanding that relatives will take most responsibility for caring for the elderly. The relationship between informal and formal care provision, and whether they substitute for or complement each other, has been the subject of intense debate and ongoing research (e.g. Adamek, 1992; Chappell and Blandford, 1991; Edelman and Hughes, 1990; Greene, 1983; Larragy, 1993; Litwak, 1985; Litwak *et al.*, 1990; Lyons and Zarit, 1999; Noelker and Bass, 1989; Twigg and Atkin, 1994; Ward-Griffin and Marshall, 2003).

Attitudes to the role of family relative to the state and the market in the provision of home-based care to frail elderly people vary widely between cultures, places and times. For example, in Sweden the assumption is that the state will provide most care for the elderly (Johansson, 1991; Sundström and Tortosa, 1999), whereas in Japan and Korea the assumption is that family will take primary responsibility (Harris and Long, 1999; Lee and Sung, 1998). Even this is subject to change: in Sweden, cutbacks to services in the 1980s and 1990s have led commentators to argue the state is placing responsibility on families to provide care; in Japan, a mandatory long-term care social insurance system with the goal of the 'socialisation' or state assumption of care for the elderly introduced in 2000 now caters to around 10% of the population aged 65 and over (Creighton Campbell and Ikegami, 2003). In some countries, the state provides funding either to care recipients or to 'informal' care-givers to provide care, thus blurring commonly held perceptions of the distinction between formal and informal, paid and unpaid care. For example, in Britain, in Austria, in Denmark, in the United States and in the Netherlands, there are various forms of provision for the compensation of care work given by families or other informal providers, although many of these programmes are seen as either controversial or naturally limited (England and Linsk, 1989; England *et al.*, 1990; Horl, 1993; McDaniel, 2000). The Direct Payment scheme in Britain, whereby local authorities give funding directly to disabled people, parent carers and older people to give them the financial ability to pay for their own personal assistants, in place of the 'social' services previously provided by the local authorities, is one example of this (Pickard *et al.*, 2003). In the United States the market is seen as being responsible for the provision of home care for elderly people, although in actuality this requires a significant amount of policy-based and direct financial support from the state (see e.g. Scully *et al.*, 2003).

There is also an interesting range of literature which, rather than modelling direct relationships between informal and formal care-givers, examines the relative roles of the state, society and family in taking responsibility for the payment for and provision of care of the elderly. Much of this is done through critical comparative analyses of national approaches to care-giving, and connects these to issues of gender inequality and citizenship in the context of the restructuring of health and social welfare systems (Glendinning, 1998; Glendinning *et al.*, 2000; 2001; Glendinning *et al.*, 1997; Hutten and Kerkstra, 1996; Knijn and Kremer, 1997; McDaniel, 2000; Ungerson, 1997). This question of the role of family in a society, particularly with respect to the community-based care of elderly people, is thorny. Qureshi points out that there are differences between statistical norms (what most people do) and normative beliefs (what most people think is the right thing to do) (Qureshi, 1990). Just because families provide most care to elderly people at home does not mean that this is what most people believe is the appropriate way to care for the elderly.

In the home setting relationships between professional and informal carers are unique. Because care for elderly people in the home setting is often long-term and gradually increasing in intensity, the professional expertise of nurses is balanced against the intimate personal knowledge of family care-givers (McKeever, 1999). This is sometimes characterised as the difference between 'caring for', on the part of formal care-givers, and 'caring about', on the part of informal care-givers (Graham, 1983). Different groups (family care-givers, seniors, care managers, care workers, other health professionals) may have very different ideas about the most appropriate boundaries and relationships between formal and informal care-givers, and the respective strengths (and hence value and recognition) of lay and professional care and the degree to which it is appropriate that family care-givers can and should take on various tasks (Pickard *et al.*, 2003).

Continuity in formal support is important. That is, formal support providers who are able to establish a long-term relationship with a care receiver may be better able to 'read' that elderly person and their needs (for example, when they are beginning to tire), and can act as an advocate for their health needs. As the work is more long term, these close socio-professional relationships may be an important form of support between care workers and family care-givers (Bass *et al.*, 1996; McKeever, 1999).

The rapid and ongoing nature of much of the provision of home care for elderly people also has an impact on relationships between formal/paid and informal/lay care-givers. As responsibility and control for the provision of home care has been decentralised, there is more spatial variability in the nature of service provision. For example, there may be wide variations in levels of public funding between regions or even municipalities (Morris *et al.*, 1999). Moreover, different local governments and different sectors such as voluntary and private or for-profit organisations interpret and implement central government policies within localised contexts. Milligan's research on the changing context of home care in Scotland provides a strong illustration of the differing social (informal) and support (formal) networks available to care-givers as a result of their spatial location relative to the changing concentration and nature of formal support services (Milligan, 2000; 2001).

As more care is provided and consumed at home, this diversity and continuous uncertainty of service provision across communities leads to confusion. Lack of knowledge about services on the part of users leads to lack of use and an inability to participate in debates about cuts and changes to services by those most affected by them (Cloutier-Fisher and Joseph, 2000). On a day-to-day level, the lack of clarity and detailed information about who performs which tasks, in which contexts, has a significant impact on informing the interaction between professionals and 'lay' carers. This influences the types of service which are offered to support informal care-givers, which in many contexts is negotiated in face-to-face encounters, thus adding to the vulnerability of informal care-givers (Pickard *et al.*, 2003).

These negotiations often take place at a time when care-givers are least equipped to deal with them, for example following a health crisis on the part of the care receiver or a time when the care-giver has reached the point where they no longer feel able to cope with the situation and is under a great deal of stress (Wiles, 2003b). As Pickard *et al.* (2003) point out, many older care-givers have been gradually taking on the role of carer over many years, and only come into contact with support services when a crisis occurs. They report that in recalling these encounters there is often a great deal of confusion on the part of carers over the role of professionals and uncertainty as to how the plans and decisions that were made were reached. Informal care-givers in a Canadian context reported similar experiences, pointing out that most people have very little sense of how the formal support system works until they need it. Acquiring support services was often reported as a disturbingly 'hit-and-miss' process of finding out information about services and eligibility criteria, and learning how to best negotiate for support without being either too passive or aggressive (Wiles, 2003b). All these factors mean that informal care-givers tend to take on greater responsibility for care.

Private informalisation of care is taking place at home . . .

Glazer (1990) showed that a shift of highly technical health work is taking place, as work previously done by paid health-service workers in health facilities is transferred to women family workers who do it for free. She points out that the highly technical work done by family care-givers challenges conventional divisions of the social world and the provision of health care into public and private. Instead, earlier discharge of patients, treatment in outpatient clinics and more home-based care mean that the free labour (usually of women) substitutes for the once-paid labour of nursing personnel (also usually women).

As services are being shifted from more highly trained health professionals to less qualified personnel or even untrained lay groups, we are seeing what might be called a 'private informalisation', a quiet deprofessionalisation of care (Wiles and Rosenberg *in submission*). Services previously provided by registered nurse practitioners may now be provided by licenced nurse practitioners, home helpers, or even family. This causes stress and the possibility of mistakes and even injury to untrained workers. Paid home helpers may receive very little training, particularly given that difficult working conditions, 'competitive' poor wages and lack of professional associations or unions lead to high turnover of staff.

The changing political–institutional frameworks and funding for home support, and the process of negotiation that takes place in receiving formal support at home, exacerbate the lack of clarity and knowledge about the professional–lay boundary. Many family or lay carers are now performing tasks that were formerly within the remit of nurses and other health professionals (Glendinning *et al.*, 2000). For example, in a study comparing three different groups of carers, including older carers of older people, Pickard *et al.* (2003: 85) show that informal or lay carers are carrying out 'not only personal and physical care tasks, but also . . . many highly complex and technical health-care tasks'. Tasks performed by older carers in this study, often unsupervised, included not only giving medication and changing dressings, but also renal dialysis, colostomy and fistula care, and parenteral feeding. Many of these tasks are highly technical and risky in terms of the potential for infection and complications.

. . . and so is an informal privatisation

Recent times have seen greater reliance on family care of the elderly (Ward-Griffin and Marshall, 2003), to the point where some commentators argue that caring is being reconceptualised by governments as a 'private' problem to be attended to by individuals and their families (Joseph and Martin-Matthews, 1993). In many countries, like Canada, the publicly funded long-term care system (comprising long-term care facilities, and community services) is shadowed by a system of informal care provided by family, friends, neighbours and local institutions (Cloutier-Fisher and Joseph, 2000). As the home becomes a site for the care of elderly persons, and institutional care has been cut back, the impact has been that families match the decline in public services. Analysis of a Swedish survey shows a process of reverse substitution of old-age care, whereby cuts to home help and institutional care have demonstrably led to greater inputs from more families, who shoulder heavier care commitments to match the decline, but now without sharing the burden with the state through the use of formal support (Johansson *et al.*, 2003).

In the current context of health care reform, budget deficits and rationing of publicly funded services, costs of and responsibility for care are pushed onto families and other informal care-givers (Coyte and McKeever, 2001). A general trend is that elderly people at home who can get help from their families are less likely to get publicly funded services than those who live alone, especially where resources for home care provision are limited (McKeganey, 1989; Sundström and Tortosa, 1999). This is often regardless of the willingness or appropriateness of family members to provide care, and in spite of evidence that informal support is a condition for the success of formal support, rather than being a substitute (Frederiks *et al.*, 1990).

Measuring the actual costs of home care is difficult, in terms of what to measure and how to measure it (especially given the holistic nature of much home care), or to whom the costs fall (Kane, 1999). Scully *et al.* (2003) estimate that home health agencies accounted for $10 billion of the Medicare budget and more than $7 billion of the Medicaid budget in 2002, and in that same year the HHC industry accounted for a total of $33.2 billion while home-based respiratory and infusion services accounted for a further $9 billion. There is evidence to suggest that a significant proportion of the costs of the provision of home care to elderly persons is borne by elderly persons themselves and their families (see below).

There are individual and household financial costs involved in the provision and consumption of care at home also. Studies in both Canada and Britain suggest that almost half of family care-givers pay out-of-pocket expenses related to care, a significant proportion of these in excess of $300 per month (Bamford *et al.*, 1998; Health Canada, 2001a). A study commissioned by the Canadian

Research Institute for the Advancement of Women points out that when care takes place at home, many costs that would be absorbed by the government (such as special meals, medical equipment, renovations to accommodate disabilities, repairs and maintenance of the care setting, and sometimes prescription and non-prescription drugs) are paid to some extent by family members (Morris *et al.*, 1999). The out-of-pocket amount paid by households varies between jurisdictions and according to the type of care provided. For example, one United States study suggested that the average annual cost of providing informal care to a community-dwelling person with dementia, accounting for care-giving time and lost earnings, was in excess of $18,000 (Moore *et al.*, 2001).

There is no lack of recognition that the costs of care need to be shared more equitably. For example, participants at a Canadian national conference on home care agreed that acute care restructuring and downsizing have placed a 'non-sustainable burden of care on individuals and on families', and that a national, government-supported and funded home care programme is urgently needed (Health Transition Fund, 1998). A Canadian public opinion poll released at the time suggested there is more popular support for investing in a home care programme than in other health system initiatives.

Care of the elderly at home is a gendered phenomenon/experience

A huge range of research demonstrates the gendered nature of the provision and consumption of care for the elderly at home, with women being more likely to both provide and receive care (e.g. Finch and Groves, 1982; McDaniel, 2000; Miller and Cafasso, 1992; Montgomery and Datwyler, 1990; Rose and Bruce, 1995; Rose-Rego *et al.*, 1998; Stoller, 1994; Stone *et al.*, 1987; Williams, 2002; Young and Kahana, 1989). As the majority of home care consumers and providers, women are most affected by reforms to those services (Armstrong and Armstrong, 2002; 1999; Armstrong and Kits, 2001; Morris *et al.*, 1999; Older Women's Network and Registered Nurses Association of Ontario, 1998). For example, differentials in life expectancy mean that women are less likely than men to be able to rely on the care of their partner and thus o rely more on professional support, which means they are more vulnerable to reductions in state pensions and reforms to health and welfare services (Arber and Ginn, 1995; Scott and Wenger, 1995). Yet women's roles as providers and consumers of care at home are often 'hidden' or invisible, especially because as a supposedly 'private' sphere activities taking place in the home are often overlooked (Angus, 1994; Graham, 1985; McDaniel, 2000). Women's roles as unpaid, and underpaid, care-givers already contribute to the income gap between men and women. Therefore inadequate funding for home care services (which leads to barriers to access to support), expectations that family members contribute to care without pay or with very low pay and at considerable cost to other aspects of their lives (health, career, pension accumulation, benefits, income), and poor working conditions for paid home care workers all contribute to women's vulnerability to poverty (Morris *et al.*, 1999).

Gender is a 'complex social location that includes race, ethnicity, class, and sexual orientation', among other things (Lorber, 1990). Other aspects of social difference interact with gender to shape the experience of care. The work of Sara Arber and others has been helpful in theorising the interaction of gender, age and caring. They challenge stereotypes of elderly people as dependent and a burden, illustrating the importance of a theoretical approach to understanding the socially constructed nature of ageing (e.g. through theorising the gendered nature of age), and revealing the importance of structural inequalities related to class and gender. For example, analysis of a survey of care work in Britain shows that the largest caring contribution is made by women in their 60s, that older men are more likely than younger men to provide care, and co-resident spouses spend the same amount of time caring whether they are male or female (Arber and Ginn, 1990). They argue that marked gender differences are more visible among younger people and those providing care to someone in another household. They later report that elderly women report worse self-assessed health than men, and this shows a strong health gradient (Arber and Ginn, 1993). Health inequalities in later life are strongly associated with previous individual occupation as well as with wider aspects of life such as the material conditions of a household and particularly with disadvantaged material circumstances (Arber, 1997).

Marriage is another factor in the use of health services. For example, an American study shows that married elderly people consistently use higher-quality hospitals and have shorter lengths of stay (Iwashyna and Christakis, 2003). These authors suggest that marriage provides elderly people with the resources to change the approach to achieving health goals, such as time to research differences in health services and how to use the best services, broader social networks offering better access to information and resources, and greater help in getting the most desired kinds of care (Iwashyna and Christakis, 2003).

The cultural sensitivity and appropriateness of home care services is also crucial to their accessibility and use. A Canadian study shows that lack of cultural sensitivity leads to under-use of home care services by Aboriginal women, and some ethnic, racial and linguistic minorities, meaning that unpaid women provide all the home care in these communities (Morris *et al.*, 1999). Lack of access is exacerbated by assumptions that care of older native people by kin is traditional and therefore not in need of support, leading to a lack of support for their care-givers, most of whom are women (Buchignani and Armstrong, 1999). A government-funded study in Canada shows that 15% of those caring for a family member at home spoke a first language other than the two official languages of the country, thus presenting a potential linguistic barrier to accessibility of care for a significant proportion of the population (Health Canada, 2002).

Employment also has an impact on informal care of elders, while gendered geographies of work and family life impact the ability to balance work and family commitments (Cloutier-Fisher and Joseph, 2000; Hallman and Joseph, 1997; 1999; Joseph and Hallman, 1996; 1998; Joseph and Martin-Matthews, 1993; 1994). One concern with respect to the ageing population and care at home is that with increasing numbers of middle-aged women in the paid workforce, the 'traditional carers' are no longer so available or willing to provide care (Doty *et al.*, 1998). For example, researchers in Japan note that decades of low fertility have meant that women are now a focus of policy, as a means both to compensate for declining numbers of male workers and to care for the rapidly ageing population (Ogawa *et al.*, 2003). In contrast, Milligan argues that community-based care policy attentive to the needs of informal care-givers must take into account changing patterns of labour and inter-regional variations in demography, which affect who needs care and who is able to provide formal and informal care. For example, the feminisation of the workforce, more geographical mobility and changing family structures mitigate against statutory expectations of an increase in the caring role of women (Milligan, 2000: 51). She notes, however, that these same factors combined with the increasing life expectancy of men might contribute to a future rise in the number of men undertaking the care-giving role.

In practice, employment has tended to mean that women juggle more responsibilities and more work, or withdraw from the paid workforce, rather than substituting work for caring (Joseph and Hallman, 1998). A recent study of African-American care-givers, a group traditionally stereotyped in the literature as relying more on informal support than care-givers in the 'white' population, shows that employed care-givers do not provide significantly less care than unemployed care-givers, nor are they more likely to use formal services. On the contrary, care-giving is more likely to influence employment, with unemployed care-givers remaining unemployed because of their care-giving duties (Bullock *et al.*, 2003). A US national study of care-givers showed that 9% of care-givers reported they had quit their jobs because of care-giving, while another 20% reported conflict, and that women were much more likely than men to withdraw from paid employment because of their caring responsibilities (Stone *et al.*, 1987).

Matthews and Campbell, in a study of the experiences of employed care-givers, show that although male employees may identify as care-givers, in practice the tasks they perform, the time they spend and the responsibility they assume tend to be less than that of women employees, and that the experiences and consequences of employees providing care to an elder is profoundly gendered (Matthews and Campbell, 1995). Yet more men are providing care at home, and it seems that they may experience different difficulties than do women (Arber and Gilbert, 1989; Cliff, 1993; Harris, 1993; 1998; Harris and Long, 1999). For example, they may have to deal with the negative perceptions of others about their role and ability.

Even the relative importance of formal and informal support is gendered. For example, a study in Manitoba, Canada shows that women are more likely to use informal support, but that over time men were less likely to use informal support and more likely to use formal support (Chipperfield, 1994). Chipperfield suggests this is because formal support is increasingly available and acceptable, and men have less access to informal support networks. Others argue that married women are more likely to receive formal support than married men, but this pattern does not hold for unmarried persons (Mutchler and Bullers, 1994). That is, in terms of informal care-givers, married women are less likely to get formal support in caring for their spouse than a married man to care for his spouse.

In contrast, a British study demonstrates that, while it is hard to tell whether men get more help from formal services than women because little is known about the most 'service-resistant or service-deprived' people, caring tends to be represented as normal and natural to women, but something special when performed by men (Rose and Bruce, 1995). Rose and Bruce's qualitative interview-based study of older male and female spousal care-givers suggests that while men and women performed similar tasks and often expressed similar feelings of sadness, frustration, resignation and anger, men were at an advantage in that merely to attempt to care was seen as admirable in a man, whereas for women it was seen as a natural duty and an obligation. They argue that the 'differential esteem felt by and for the older carer speaks eloquently of the long and deforming fingers of gender' (Rose and Bruce, 1995: 127).

As a new site of care provision, the growing importance of home care has had significant impacts on the predominantly female paid labour force of home care workers. Health visitors to the home tend to be women, and some authors speculate that this is because women tend to be socialised into particular roles (Rolls, 1992). On first glance, greater potential for home-based care work might offer good opportunities for female workers in that it is a job with potentially flexible hours and relative autonomy. For example, early home care programmes in Sweden were primarily staffed by previously unemployed housewives, largely to the satisfaction of both employees and care recipients, who reported the arrangement to be a friendly and relaxed one for all (Sundström and Tortosa, 1999). More recently, however, in the context of limited budgetary resources and increasing demand, the opposite has largely been the case. Restructuring of policy and home care work has had serious negative consequences for the predominantly female home care labour force (Williams, 2001). Moreover, health professionals working in homes are paid considerably less than those working in institutions (Morris *et al.*, 1999). Poor working conditions, wages, high turnover, safety issues and health issues have severe negative impacts on employees in the home care industry, and the majority of those who are negatively impacted are women (Morris *et al.*, 1999; Williams *et al.*, 2002).

Care at home is shaped by context/proximity

In an influential study, Qureshi and Walker illustrate that the likelihood of who provides care to an elderly person can be modelled according to a hierarchy shaped by relationship and gender, showing that responsibility tends to fall first to a spouse, or an adult daughter, adult son or daughter-in-law (Qureshi and Walker, 1989). Later work shows that this hierarchy of informal providers still holds (Bond *et al.*, 1999).

Yet some researchers have argued that it is living arrangement, rather than marital status per se, that determines the likelihood of receiving support, and that living arrangements also shape who is likely to be the primary care-giver (Chappell, 1991). When available, a spouse is the most likely source of assistance, followed by children. But, in findings contradictory to those of the hierarchical support model, Chappell (1991) demonstrates that if neither of these are available, the next source of support is likely to be sought from friends for those who live alone, and siblings for those who live with non-spousal others.

Others have pointed out that Qureshi and Walker do recognise that co-residence overrules hierarchical relationship, and is in fact highly influential in defining who within a family ends up as the carer, overriding factors such as gender and relationship (Twigg and Atkin, 1994). Several other researchers have shown that proximity shapes the likelihood of informal care provision, and what kinds of care are provided. Closer proximity (such as co-residence) increases the likelihood of informal provision of personal care such as dressing and bathing, and the amount of time spent caring (Arber and Ginn, 1990; Glendinning, 1992; Moore *et al.*, 1997).

Indeed proximity itself, and even co-residence, may be a matter of decision-making and migration patterns related to potential need for care, rather than simply a predictor of providing care. For example, one study shows how decisions about proximity are often based on a range of considerations including potential care-giver time, anticipated difficulties in shared households, relationship to the elder, and the ability of the elder to reciprocate by providing assistance to the carer (Stoller, 1988). Co-residence with an elderly care recipient has been demonstrated to be a complex discriminator of work–family balance, so that distance between carer and cared-for is as important as or more important than distance between home and work (Joseph and Hallman, 1996: 395). Yet the benefits of cohabitation in terms of reducing the interference of work or work-related stress may be offset by having dual responsibility (e.g. for children as well), and also living arrangements often change very quickly in response to greater stress levels (e.g. encouraging the elderly recipient to move closer).

Conclusion

Demand for and provision of home care for frail elderly people is growing, due to perceptions about the therapeutic and social benefits of consuming and providing care at home, to changing social and demographic characteristics of populations, to developments in technologies and approaches to health and health care, and to fiscal imperatives to reduce health care costs. As societies struggle to meet this challenge, the rapidly changing political and institutional frameworks within which home care is provided and consumed present challenges in and of themselves for the range of actors involved, including not just elderly people but also those who care for them on an informal or lay basis, and on a formal, paid, for-profit and not-for-profit basis.

Greater political and public interest in the idea of 'ageing in place' suggests that we need a critical engagement with the idea of the home as a social and spatial context for ageing. Homes are not simply neutral containers which pre-exist as ideal places for the provision and consumption of care for the elderly, but require a good deal of physical, social and emotional work to become and to be maintained as sites for care.

Even then, the lived experience of care at home does not always match up to ideals. In the context of care at home, and in the context of continuous restructuring, relationships between formal and informal carers take on new meanings. Research points to disturbing trends taking place even within societies that subscribe to a collective responsibility for the provision of health care services. Perhaps because homes have traditionally been conceptualised as belonging to the private sphere, the shift of responsibility for the provision of home care from the state to other actors, particularly unpaid and lay caregivers, is less visible and less subject to contestation and resistance than cuts in more public spheres of institutional spending such as hospital closures. Consequently, we see that a 'private informalisation' of home care is taking place, and so is an informal privatisation. Most care for the elderly at home is provided on an unpaid, lay basis, and decisions about who provides care are not always made by choice so much as by crisis and necessity. The consequences of these shifts are not equitably distributed across the population. Care of the elderly at home is a gendered phenomenon and experience, and women bear the brunt of the costs and responsibilities for care at home, both as providers and as consumers.

7

HEALTHY AGEING IN THE COMMUNITY

Helen Bartlett and Nancye Peel

Introduction

Older people have received little policy attention in relation to health promotion until the last decade. With the worldwide demographic shift to ageing of the population, policies supporting healthy, productive and independent ageing now represent an important and attractive initiative for governments (Commonwealth Department of Health and Ageing, 2002; New Zealand Ministry of Social Policy, 2001; Secretary of State for Health, 1999; US Department of Health and Human Services, 2000). The 'healthy ageing' construct has become a cornerstone of policy development in response to population ageing and this embraces concerns about quality of life as well as length of life.

While the physical dimensions of healthy ageing are understood, the concept of healthy ageing requires the adoption of a broader definition of health that extends beyond physical health and the provision of health care services, into policies to provide accessible and supportive living environments. This will require a greater awareness of ageing issues and of the heterogeneity of older people, a breaking down of stereotypes and a recognition of the positive aspects of ageing. The current culture of old age pays less attention to older people as productive members of society and the contribution of this to healthy ageing.

Given the broader context in which healthy ageing now needs to be considered, individualised approaches to promoting healthy ageing are unlikely to be successful on their own. The challenge is therefore to develop effective health promotion strategies and activities for older people involving community-based approaches. Despite the fact that community development has been less popular as an approach for health promotion in recent years (Bernard, 2000), its effectiveness as a tool for hearing the voices of older people is well documented (Beresford and Croft, 1995). A settings approach for promoting healthy ageing is therefore needed if policy goals are to be achieved.

This chapter explores the concept of healthy ageing and the issues of definition and measurement. The perspectives of older people themselves are considered in the context of formulating healthy ageing strategies for the community. Progress towards the policy goals of healthy ageing, increasingly endorsed around the world, is considered by drawing on UK and Australian examples that are guided by a settings approach to health promotion. The challenges and opportunities to maximise health in later life are considered in the broader community context.

Health trends

The health consequences of increased life expectancy are considerable. In Australia women aged 65 in 2001 could expect to live to 85.2 years and men to 81.6 years. Overall, the burden of disease and disability falls disproportionately in older age with 28% of total years of life lost to disability occurring in the older population, who account for 12% of the population (AIHW, 2002). Dementia was the leading cause of years of life lost to disability for people aged 65 and over (17%), followed by adult-onset hearing loss (8%) and stroke (7%). Vision disorders, osteoarthritis and coronary heart disease each accounted for 6% of the disability burden. Risk factors contributing to loss of healthy life years among older people include tobacco smoking, high blood pressure, physical inactivity, high blood cholesterol and inadequate intake of fruit and vegetables.

There are clearly improvements to be made in the general health status of older people if the risk factors can be reduced. For example, only 54% of older Australians exercised at the recommended levels to achieve health benefits and on current trends (AIHW, 2001), by the year 2025, approximately 2.5 million Australians over the age of 65 will be sedentary. Physical inactivity has been linked to many health conditions including type-2 diabetes, heart disease, musculoskeletal disorders, some cancers, high blood pressure, high blood cholesterol and atherosclerosis, as well as contributing to excess body weight and risk of falls. The prevalence of overweight people is also higher among older than younger Australians. Some 69% of women aged between 55 and 74 and 73% of men aged between 45 and 74 are overweight. This contributes to higher morbidity and mortality rates for type-2 diabetes, coronary heart disease, respiratory disease, some cancers, gall bladder disease, osteoporosis and ischaemic stroke.

The concept of healthy ageing

Since the 1960s gerontologists have been developing theoretical frameworks to describe ideal outcomes of the ageing process, yet the concept of healthy ageing has only recently been promoted as a guiding theme for gerontological research and as a challenge for the design of health and social policies.

However, there is confusion as to the nature and terminology of the concept. Quality of life in older people has been variously conceptualised as 'successful' (Rowe and Kahn, 1997), 'active' (World Health Organization, 2002), 'robust' (Garfein and Herzog, 1995), 'productive' (Kerschner and Pegues, 1998) and 'positive' ageing (Bowling, 1993). While American research favours the term 'successful ageing', success in Western culture is usually associated with economic achievement, employment status, income and assets. Relative success in life is also a value-laden judgement, rooted primarily in middle-class values and typically white, male standards (Baltes and Carstensen, 1996). The term 'healthy ageing' favoured by Australian researchers, stakeholders and policy-makers (Kendig *et al.*, 2001) is in common usage in Australasia and Europe (Hermanova, 1995) but also appears to underlie the health policy development in the USA (US Department of Health and Human Services, 2000) and Canada (Health Canada, 2001b). The difficulty in advancing healthy ageing lies in the nature of the concept itself. There appear to be no standards or underlying theme for defining healthy ageing outcomes.

The concept of healthy ageing recognises that health, as defined by the World Health Organization (World Health Organization, 1952), extends beyond the mere absence of disease or infirmity to include physical, mental and social well-being. Ageing is seen as a maturation process of development and change over the life course. The Ottawa Charter for Health Promotion (World Health Organization, 1986) provides a conceptual framework that recognises that broader environmental and social factors, and not merely individual behaviours, are key determinants of health status and thus the prospects for healthy ageing (Minkler *et al.*, 2000). Incorporating these elements into a definition, healthy ageing is a process of adaptation to physical and psycho-social changes across the life course to attain optimal physical, mental and social well-being in old age.

Older people's perspectives

To reflect the realities of ageing, the most appropriate people to define what healthy ageing means are older people themselves. For them, health is an important quality-of-life dimension (Stewart and King, 1994). Of all the conditions studied in connection with ageing well, good health, broadly defined, is the criterion most generally agreed upon (Day, 1991) However, the meaning of health in old age shows large cultural, gender and age differences and will be interpreted in relation to chances of healthy ageing. Subjective values about what constitutes a 'good old age' are also relative to time and place, involving images about old age and the social roles appropriate in later years (Day, 1991). Because personal meanings of health relate to conditions that pertain at a particular point in history and to the particular setting of economic and political priorities, the relativity to time and place suggests that the current meaning of healthy ageing may be outdated for succeeding cohorts (Day, 1991).

People's perceptions of their own health have consistently emerged as a significant indicator of current and future well-being (O'Rourke *et al.*, 2000). A growing body of evidence shows that older people perceive themselves as healthy even in the face of real physical problems (Rowe and Kahn, 1999). This suggests that, with ageing, people adapt to changes in their objective health and functional performance and tend to assess their health as the same, or even better, with increasing age despite an increase in chronic diseases and decline in functional performance (Leinonen *et al.*, 2001).

A number of studies in Australia (Earle, 1996; Kendig *et al.*, 1996), Europe (Baltes and Mayer, 1999; Grundy and Bowling, 1999; von Faber *et al.*, 2001) and the USA (Bryant *et al.*, 2001; Day, 1991, Kerschner and Pegues, 1998) qualitatively explored the views of older people on the meaning of, and ways to achieve, healthy or successful ageing. Despite limitations of generalisability across cultures and cohorts, a number of consistent themes emerge from these studies. For the individual, healthy ageing means having a sense of well-being, the capacity for independent activity, meaningful involvement, supportive environments and positive attitudes. Being healthy is seen as having the resources for an everyday life that is satisfying to self and others.

Research perspectives

From a research perspective, definitions of healthy ageing are focused on measurable outcomes. To direct research, theoretical models of healthy ageing have been advanced to identify when and why health may be labelled as optimal, to understand the determinants, and to suggest ways to modify the nature of human ageing, if desirable.

Biomedical researchers distinguish between healthy ageing and pathological ageing in terms of effects on mortality, morbidity and functioning and thus have defined healthy ageing as prolonged life expectancy free of disease and dysfunction. Biomedical models are designed to help identify, prevent and reverse functional losses associated with usual ageing.

Fries' theory of compression of morbidity (Fries, 1980) is an example of the biomedical model of healthy ageing aimed at extending the years of healthy life by postponing or eliminating causes of morbidity. Healthy ageing consists of optimising life expectancy while at the same time minimising physical, psychological and social morbidity overwhelmingly concentrated in the final years of life. The model suggests that the physiological and psychological variables which predict future ill health would be identified and modified through specific interventions involving alterations in lifestyle, the environment, or both (Fries, 1996). The model, however, focuses on the causes of age-associated deficits. Interest in healthy ageing requires a paradigm shift to focus on those persons who are ageing well.

A paradigm shift towards positive (healthy) aspects of ageing was introduced in 1987 by Rowe and Kahn in the context of separating the effects of disease from the ageing process itself. The model proposed that the 'normal' non-diseased population can be stratified into two major groups identified as 'usually' ageing and 'successfully' ageing individuals. In this model, successful ageing was defined as multi-dimensional, encompassing the avoidance of disease and disability, the maintenance of high physical and cognitive function and sustained engagement in social and productive activities. Studies have been designed to show the efficacy of modifications of environment and lifestyle factors for increasing the likelihood that older individuals might achieve success under this triarchic definition.

The biomedical models emphasise health outcomes such as staying out of institutions, not being reliant on continuing input from service providers, and remaining mobile and competent in activities of daily living, based on the assumption that people have to achieve certain physical and mental goals to be considered as ageing well (Bowling, 1993). Functional loss, however, does not necessarily equate with the capacity to adapt and live normally (Lawton, 1991).

Other models of healthy ageing have emphasised the social and psychological environment of ageing and are presented as alternatives to the 'loss-deficit' model. Sociologists see healthy ageing as encompassing areas including quality of life, well-being, work, independence, self-care, financial control and management (Kendig *et al.*, 2001). Psychological perspectives define positive ageing as attainment of life satisfaction, coping behaviours and retention of social support networks (Bowling, 1993). These perspectives emphasise the process of adapting to the interactions of ageing, changing health status, self, family and society, centred on feelings of control.

An example of a psycho-social theory which elaborates on adaptive behaviour for healthy ageing has been proposed by Baltes and Baltes (1990). In this model, successful ageing is seen as an adaptive process involving components of selection, optimisation and compensation (SOC) in a search for the positive aspects of ageing. Individuals facing changes in abilities as they age can adapt through selecting activities, optimising skills and compensating for losses. While selective optimisation with compensation is considered a lifelong phenomenon, it is amplified in old age because of the loss of biological, mental and social reserves. Successful ageing is defined as the maximisation and attainment of positive (desired) outcomes and the minimisation and avoidance of negative outcomes. The model suggests potential strategies for successful ageing including:

- engaging in healthy lifestyles in order to reduce the possibility of pathological ageing;
- encouraging individual and societal diversity;
- strengthening reserve capacities (physical, mental and social) via educational, motivational and health-related activities and nurturance of social convoys;
- provision of societal resources and social opportunities; and
- facilitation of adjustment to reality.

A key outcome of the Valencia Forum, the scientific meeting preceding the second World Congress on Ageing in 2002, was the identification of a policy research agenda. One of the major priorities was identified as the need to define and delimit the scope of healthy ageing and understand its determinants. Until we have such research evidence, greater caution should be exercised by policy-makers in the use of the term 'healthy ageing' and claims that it is an achievable goal.

Policy perspectives

There is a noticeable shift internationally in the policy agenda on ageing from a focus on aged care to the broader health, social, psychological and economic concerns. The concept of healthy ageing was first defined by the World Health Organization in the UN Plan of Action on Ageing, drafted by the 1982 United Nations World Assembly on Ageing. Subsequently the World Health Organization developed *Active Ageing*, a policy framework 'to optimize opportunities for health, participation and security in order to enhance quality of life as people age' (World Health Organization, 2002). The WHO active ageing policy framework sets out what is known about the determinants of active ageing. It is guided by the UN Principles for Older People (independence, participation, care, self-fulfillment, dignity) and requires actions on three basic pillars: health, participation and security. The WHO (2002) recognises the challenges of achieving active ageing, in particular the 'double burden of disease' experienced by developing countries, the increased risk of disability with ageing, providing care for ageing populations, the feminisation of ageing, ethical issues and inequalities, and the economic costs associated with ageing populations.

Many countries have developed their own national strategies to address population ageing. The *National Strategy for an Ageing Australia* (Commonwealth Department of Health and Ageing, 2002) offers a broad-ranging framework to identify the challenges and possible responses for government, business, the community and individuals to respond to the health, social and economic challenges of Australia's ageing population. The strategy sees healthy ageing as requiring individual, community and public and private sector approaches to maintain and improve the physical, emotional and mental well-being of older people. Recognising the intertwined relationship of the community environment with individual health, the strategy extends beyond the health and community services sectors, since the well-being of older people is affected by many different factors including socio-economic status, family and broader social interactions, employment, housing and transport (Kendig *et al.*, 2001). Priority areas have been further set out in the *Commonwealth, States and Territories Strategy on Healthy Ageing* (Healthy Ageing Task Force, 2000).

Healthy ageing is also the subject of a report prepared by the Australian Prime Minister's Science, Engineering and Innovation Council (2003). It sets out a vision for an additional ten years of healthy and productive life expectancy by 2050 and recommends health promotion and prevention action in physical activity, mental activity, nutrition and pharmacological interventions. In addition, the need for an age-friendly society is recognised – one in which work and social environments, and the built environment, are more inclusive of an older population. The report highlights the need for older people to maximise their mobility and independence, to actively participate in society, engage in continued lifelong learning and receive high-quality health care (PMSEIC, 2003).

Healthy ageing through a settings-based approach

While there is still more to understand about the determinants of healthy ageing, there is clear evidence of the need for it to be promoted within a community setting that provides the context and social structure for effective action (Poland *et al.*, 2000). Research has confirmed the health-protecting effects of social support and social networks on the health and well-being of older people (Findlay, 2003). This requires a major shift in how health is conceived, replacing the dominant disease and medical model which focuses on individual change, with a population approach that is concerned with the settings in which people live, as articulated by Ashton and Seymour (1988) in the 1980s. The community development model offers a guiding framework for healthy ageing. It aims to create active participation of individuals, address the perceived needs of community, promote empowerment of individuals and community and achieve an equitable distribution of power and resources. By tackling local issues, this approach seeks improvement of community infrastructure such as housing, employment and education. In recognising the broader socio-economic determinants of health, health promotion approaches are returning to community-based models. The importance of adopting a more holistic approach that utilises social and personal resources is recognised in healthy public policy strategies such as the settings approach. The development of this approach is only recently reflected in health policies concerning older people.

More still remains to be achieved in terms of building social capital and reducing inequalities, as the measures of progress are still largely disease-based (Davies, 2001). Challenges also exist in developing new partnerships that move health promotion into other settings beyond the traditional health sector. In Australia, some of the limitations of the Healthy Cities movement were found to be creating partnerships and engaging local people in policy decision-making about health (Baum, 1993). While the policy rhetoric about healthy ageing acknowledges the broader societal context, there will be inevitable challenges in the translation of policy into practice.

Translating theory and policy into practice

A range of healthy ageing initiatives in the community have been developed in recent years, initiated by government, charitable organisations or local community groups. These include support groups, falls prevention, seniors' fitness classes, walking schemes and other physical activity groups. However, evidence about their effectiveness is mainly anecdotal and few evaluations have been published. Some examples of initiatives that have been evaluated or of projects under way are described below. The extent to which they embrace community-based approaches varies, but attempts to promote such concepts as partnership, community participation and empowerment are found.

Community walking schemes

Community walking schemes have received considerable attention in recent years as an effective means of physical activity for older people and such schemes lend themselves to a community development approach. One Australian example is the National Heart Foundation 'Just Walk It' programme, established in over 50 areas throughout Queensland. Through organised walks of varying levels, it aims to increase the community's participation in regular and enjoyable physical activity. Each local area has a coordinator and local working group whose roles are to support local walk organisers, to recruit walkers and to liaise directly with the Heart Foundation. An important feature of the scheme is its framework of partnerships between local community organisations, state and local government and community health centres. This model enables the program to be adapted to meet particular local needs. Evaluation of the scheme is under way, but there are promising signs that this collaborative community-based intervention may be effective in sustaining individual-level improvements in physical activity (Foreman *et al.*, 2001). While the Just Walk It programme is open to all ages, its approach has particular appeal to middle-aged and older adults.

Experiences in the UK of a similar walking scheme (Health Walks) have been positively evaluated in terms of adherence, satisfaction and take-up of other physical activities (Ashley and Bartlett, 2001; Lamb *et al.*, 2002). Key elements of the UK programme are community participation, use of volunteers, local resources (footpaths), initial support and ongoing referral from primary care. The communities in which the schemes operate have considerable ownership of the activity and this has led in a number of cases to the development of a network of related activities. Being involved in the local community was found to be a key factor motivating people to participate in the Health Walks scheme. A limiting factor of this initiative has been that the initial projects were established in predominantly white, middle-class locations. One of the challenges will therefore be engaging the interest of communities of lower socio-economic status and with more ethnically diverse populations.

Self-health promotion

A range of international examples in which the health of older people is the focus have been described by Bernard (2000). She documents a series of initiatives occurring in the UK through the 1980s and 1990s, including the 'Ageing Well UK' programme, which operated from 1994 to 1997 and focused on 'maximizing the health potential of older people as a health resource for their peers'. While such examples illustrate the development of a more holistic approach, Bernard argues that a comprehensive framework for action and research has not emerged.

The UK Beth Johnson Foundation action research initiative described by Bernard does, however, appear to offer a more successful community-based approach. It focused on four main elements: a Senior Health Shop, a peer health counselling scheme, a telephone link service and health-related courses and activities. The evaluation of this scheme identifies eight sets of lessons for practice: setting, publicity, involvement of older people, multiple sponsorship, professional inputs, training and resources, research, and evaluation and time. Perhaps the most crucial of these lessons is the need to establish initiatives 'where people are'; that is, building on settings that are familiar and comfortable to older people and recognising that there will be a wide range of settings where health promotion can occur. Fundamental to this is the need for community engagement of older people through active participation and empowerment and the use of pre-existing groups and older people's organisations. Partnerships with primary-care professionals are also vital initially.

While this project demonstrated that a community development approach works well with older people, not all the goals of the project were met. Participation of older men and black and minority ethnic elders was quite limited, raising an important question about inclusiveness. It was also noted that distinctions between the professionals and the older people remained quite obvious, in spite of the desire to work in partnership. Finally, there was an over-emphasis in the project on the individual level behaviour change model and there were difficulties in making connections with the wider social, political and economic context.

Community-oriented primary care

General medical practice is recognised as an effective platform to deliver health promotion to older people because of the regular contact, respect and trust that patients have for their GPs (Elley *et al.*, 2003). Although GPs may hold negative perceptions of older people and question the effectiveness of health promotion activities, GP interventions have been found to produce positive patient outcomes for older people, including self-rated health, increased walking and other pleasurable activities (Kerse *et al.*, 1997). However, in order to be successful and to improve levels of physical activity, arguably physical activity interventions need to be introduced as part of wide-ranging community strategies.

A community-oriented primary care approach (COPC), adopted and tested in the UK, was found to be effective by taking a broader community and public health perspective within primary care for older people (Illiffe and Lenihan, 2001). The model was used in four practices and involved a continuous process of community diagnosis, problem assessment, implementation and evaluation. Surveys and consultations within the defined communities were important to the process of prioritising the communities' issues, as was a multi-professional steering group and non-directive support from an academic department familiar with general practice. The model appeared to be successful in implementing interventions with community members using existing resources. While the project was conducted in affluent communities, the older population within these communities was considered to be relatively disadvantaged. Some of the lessons learnt from this project included the limited practical advantages of some new theoretical models, the over-ambitiousness of certain interventions and the difficulties of maintaining leadership at a practice level. The limited time available to formulate an appropriate project design following identification of the community's health problems was also recognised.

Injury prevention

Programme activity in relation to accident prevention has increasingly involved community-based approaches. Falls prevention is identified as an injury prevention area in the *National Injury Prevention Plan: Priorities for 2001–2003*. In Western Australia this stimulated a state-wide falls prevention programme for seniors (Stay On Your Feet WA programme), underpinned by the philosophy of coordination, cooperation and community involvement. This was informed by the North Coast of New South Wales Stay on Your Feet programme (Garner *et al.*, 1996), which applied health promotion strategies in the Ottawa Charter for health promotion and had received positive evaluations (van Beurden *et al.*, 1998). The aim is to reduce the expected rate of falls-related injuries in seniors living independently in the community, empowering individuals to take action to prevent falls. The focus goes beyond a concern with medications and chronic health conditions, although they are included in the strategy.

The scheme has involved seniors' organisations, service providers, local and state government and active and visible participation of older people. A key aspect of the programme is the awareness information and community involvement. Volunteers are recruited and seniors are trained as falls prevention advisers and they have been particularly successful in rural areas. The concept of learning circles is also used to engage the community. Safe home environments are another component of the programme involving work with builders and architects to reduce hazards and improve housing design. Public environments are addressed through collaboration with architects, local councils, environmental planners and builders. Recreational activities through local councils and recreational centres are promoted, e.g. walking groups, t'ai chi and yoga. Physical activity, in particular balance and gait, is a strongly promoted element of the programme and industry groups are encouraged to develop and expand the range of physical activity events for seniors. Evaluation reveals achievements in many areas, in particular the development of the peer education and volunteer service, although the difficulties of measuring the impact of the various activities on increasing awareness of Stay On Your Feet remain a challenge.

Age-friendly community standards

An important foundation for healthy ageing is the development of an 'age-friendly' or 'senior-friendly' society – one that is inclusive of older people.

There has been a recent surge in the development of age-friendly services and facilities, as evidenced by a search of websites detailing age-friendly banking, tourism, public toilets, homes, transport and businesses (Worrell *et al.*, 2002). However, there is little empirical evidence to demonstrate whether age-friendly standards do improve participation outcomes. Research projects under way in 2003 at the University of Queensland's Australasian Centre on Ageing are evaluating processes and outcomes of age-friendly standards in the case of public transport, and also exploring the factors that contribute to making new communities age-inclusive.

An age-friendly environment has been prioritised by the United Nations for the International Research Agenda on Ageing. In Australia, the need for work, social and built environments that are age-friendly has been highlighted (PMSEIC, 2003) to promote healthy ageing. Housing and the locational aspects of ageing are also strongly identified by the National Strategy for an Ageing Australia (Commonwealth Department of Health and Ageing, 2002) as they make a significant impact on older people's lives. The current policy goal of 'ageing in place' emphasises the importance of enabling older people to remain in their own homes for as long as possible if they so wish. Of course the feasibility of this depends on the nature and quality of the housing. Design of the internal environment of the home is important in terms of space, layout, safety and use of technology. Although older homeowners may have homes of poor quality, older people living in rental housing are worst off in terms of tenure and the quality of accommodation. As better-quality housing has been linked with fewer visits to the general practitioner, there are obvious health implications (Howden-Chapman *et al.*, 1999). Dimensions of the external environment are important too and include public spaces, transport, recreation and urban planning. Where the built environment is age-friendly, older people are more likely to remain independent even if their functional capacity has deteriorated.

Opportunities and challenges

While many opportunities exist for adopting a community-based approach to promoting healthy ageing, a better understanding of what works and improved dissemination of good practice is needed. Bernard argues that 'a comprehensive preventive public health policy, delivered at a community level and linked with a strategy which emphasises the promotion and maintenance of abilities in old age, as well as (but not instead of) targets for reducing the incidence of ill health and disability, is now long overdue' (Bernard, 2000: 11). Policies for healthy ageing need to take into account the settings in which people live and adopt an inter-sectoral approach to address the diverse range of factors that impact on health maintenance.

The challenges to achieving healthy ageing, however, also need to be highlighted. Baltes and Smith (2002) point out that 'healthy and successful ageing has its age limits'. The realities of chronic illness are highlighted by the Berlin

Ageing Study (Helmchen *et al.*, 1999), which found that almost half of the 90-year-olds suffered from some form of dementia. In addition, significant hardships (e.g. poverty, isolation, widowhood) are unavoidable events faced by many in later life. Further challenges relate to the difficulties in translating healthy ageing concepts to diverse cultures. Since healthy ageing standards are based predominantly on middle-class, typically white, male values (Baltes and Carstensen, 1996), in Australia the application of this concept to Aboriginal and Torres Strait Islanders is problematic. In 2001 people from culturally and linguistically diverse backgrounds comprised one in five older Australians (AIHW, 2002). Such cultural diversity still requires much greater consideration in the formulation of policies and programmes that address healthy ageing.

There will be new challenges presented by the baby boomer cohort (those born between 1946 and 1964), whose health problems reflect considerable social and lifestyle changes. Preventive action will need to be taken during the middle years if chronic illness such as heart disease and stroke are to be avoided. But community and societal issues such as changing family structure, multiple caring roles, working practices, attitudes and expectations will all need to be part of preventive health strategies. While there is a great deal of speculation about this future cohort, it is highly likely that they will be more inclined towards community involvement in decision-making and policy and more questioning of health care professionals in general.

Nevertheless, a greater understanding of the barriers and facilitators of a community approach to healthy ageing is still needed if the notions of partnership, participation and empowerment are to be adopted in the quest for healthy ageing (Bartlett, 1999). Fundamental to this has to be ongoing attempts to influence and monitor attitudes and knowledge about ageing and older people. Valuing the significant contributions made by older people to society at many levels and promoting more positive inter-generational relations are two of the key, but currently neglected, ingredients for healthy ageing in the community.

8

THE PHYSICALLY AGEING BODY AND THE USE OF SPACE

Kichu Nair

Space plays an important role in an older person's life. The phenomenon of shrinking physical and social space as people age is well known. Older people are often prisoners in their own homes and are socially isolated. Moreover, as a society, we do not organise space for older people very effectively. For example, houses are typically designed for younger people who lack serious disabilities. As people get older their own homes become less and less user-friendly and consequently they use less space. Later, if people move to residential care, their world is even more constricted, perhaps even to a single room or a single bed. In contrast, an active young grandson will have use of a comparatively vast physical and social space. Health professionals involved in the care of older people should be more proactive in the planning, reorganising and rearranging of the space around the older person. This will enable the young people of today to continue to live in the safety, comfort and privacy of their homes, when they become old or disabled.

As populations age there is a dramatic increase in the number of frail and dependent older people. Ageing caused no major issues in the past, when it was very common for older people to live with their families. The latter part of twentieth century, however, has witnessed a change and families have ceased to provide care and accommodation for their older members. In this context, older people's uses of physical and social space become even more important (Hugman, 2001).

Loosely defined, frailty is instability and risk of loss whereas disability indicates loss of function (Campbell, 1997). In spite of this, the majority of older people live at home and would like to continue to do so. Indeed, this is often their preferred option, even though they may require assistance to do this. In many cases, the home space is where they have lived for a considerable length of time and which has much emotional and physical meaning for them. Even though some home-dwelling older people may need support from families or community care agencies for activities of daily living, they may still feel they are in control of their lives. This feeling is very important for them.

In terms of care and support, to keep older people in their homes, a comprehensive assessment of their risk of disease and disabilities is required, preferably undertaken at their existing place of residence. This will give information about the home safety and interventions, if needed. In addition, such an assessment should indicate how well a person is managing with illness and disability in the physical and social space (Rubinstein, 1999). Notably, it should also provide valuable information on the social interaction, medication compliance and risk of malnutrition of the older person.

It is often a combination of impairments that makes an older person 'very' disabled. A good example is a 45-year-old stroke patient versus an 85-year-old stroke patient. Even though they both have the same disease affecting the same part of the brain, the older patient, typically with previous impairments, will be more disabled and home safety will be more of an issue for the older patient.

Another example is the issue of falls (also discussed in chapter 11), which often result from impairment and disability and lead to further impairment and disability. They are also a major cause of mortality. In fallers, 44% of men and 65% women fall inside their residence. The falls mostly occur in the bedrooms, kitchen and dining room, reflecting the frequent use of these spaces. Approximately 25% of men and 11% women, however, fall in the garden. Most of the young-old are likely to fall outside the home compared to the old-old, again reflecting the use of different spaces in their activities and different grades of mobility. The common environmental reasons for falls are loose carpets, bathtubs without rails, poor lighting, unsafe stairs and ill-fitting shoes (Masud and Morris, 2001). All these can be fixed with some simple assessment and planning.

In an Australian study on falls at home, only one in five homes was found to be hazard-free. The majority of homes contained more than one hazard and they were in the rooms where the older persons were likely to perform complex activities (Carter *et al.*, 1997). The most interesting finding was that the majority of subjects were able to identify falls as the major cause of injury and 87% were able to accurately name at least one safety measure. Meanwhile, in another study of safety at home, all homes had at least one hazard. Floor rugs and mats were the commonest followed by stopovers and steps and trailing cords (Stevens *et al.*, 2001). Notably, the interventions were associated with a small but significant reduction in risks.

Home modification should be proactive rather than reactive, to keep older people at home. This will prevent the need for admission to residential care in the long term. Unfortunately, however, there are many forms and modes of provision and all these services require careful coordination. For example, the South East Senior Housing Initiative (SESHI) in Baltimore is a joint initiative between housing, health care, social service and community services to coordinate and study these home modifications. This project assesses how technical assistance could empower older people to make better decisions about home environments (Fisher and Giloth, 1999). Such initiatives are increasingly needed to meet the demands of ageing populations. Ideally, houses should be designed for future adaptation, when people get frail or disabled. The designs should be in consultation with building industries, architects and health care professionals.

These all ensure homes are user-friendly for older people and future modifications are easier and cheaper (Frain and Carr, 1996).

Of the 12,646 falls-related deaths in 1992, approximately 10% occurred on stairs, making this the most common cause of death in this category. However, falls are likely to be under-reported (Startzell *et al.*, 2000). Incidence of stair-related injuries is likely to increase with age. Approximately 75% of falls on stairs occurred during descent. Voluntary sector older people's support and pressure groups stress that older people should use extreme caution when using stairs, should pay careful attention to steps, especially while descending and should avoid carrying objects. An older person negotiating a flight of stairs needs input from sensory organs and must detect hazards, choose the route, perceive the foot location and continuously monitor all these. Templer (1992) has described these in a 'stair behavior model' and reported that most stairway incidents occur at the top three or bottom three steps.

In order to keep older people safer in their homes and familiar surroundings, 'assistance technology in elderly care' is fortunately available now (Miskelly, 2001). Several types of alarm are available for community dwellers to raise help, provided they have cognitive and physical capacity to summon help. Video-monitoring at home enables relatives to scan the patient's location at home. Arm bands or wrist bands could detect wandering people. Fall detectors are designed to be worn around the waist and chest. They can activate services when the person has a fall and immediately trigger the alarms. 'Pressure mats' are electro-mechanical devices to detect mobility from chair to bed and vice versa. These can provide automatic communication between the older person and their carers. Movement detectors, similar to burglar alarms, can be used by carers to monitor the older person's safety at home. All these devices may keep older people safe at home.

Most recently, the emergence of disciplines like gerontechnology (a combination of gerontology and ergonomics) is assisting older people. Related technologies assist older people's safety, performance and comfort and can compensate for disabilities and impairment. It is well known that with ageing, visual acuity, contrast and depth perception are impaired. All these can be partially compensated by proper planning and design of living space. For example, to maximise safety, walls should have no glare surfaces, doors and windows should be made of opaque materials, furniture should have no sharp corners and lights should be ceiling-mounted with fluorescent light sources (Pinto *et al.*, 1997).

One of the assets which enables older people to manage well in the community or residential care is visual ability. Traditionally vision is assessed by checking for acuity. Poor vision is evident in up to 72% of community-dwelling older people. But there are other important aspects of vision, particularly for older people. Contrast sensitivity and depth perception are important determinants of postural control and falls prevention. These factors were found to be independent predictors of total sway on a compliant surface (Lord and Menz, 2000). The relationship between physical disability and visual functions was studied in a sample of 1210 community-dwelling older females. In this study contrast sensitivity was a better predictor of functional impairment with an adjusted odds ratio of 5.1 (2.0–12.9) (Dargent-Molina *et al.*, 1996).

New technologies can link older people with the community and with their families. Older people are interested in using these. Research suggests that they like to use email to maintain contact with children and grandchildren. The use of technology is even more likely when geographical distance is an issue. In an Australian study, 16% of older rural dwelling older people used the Internet for email, browsing and accessing information. What was more interesting was that non-users were also keen to learn these and cost and lack of opportunities were the main impediments (Irizarry *et al.*, 2001).

Fire-related deaths are common in the elderly but have received scant attention in the academic literature. A study based in Alabama identified fire-related fatality as highest among older people (McGwin *et al.*, 2000). In particular, fires were a major concern because of older people's poor mobility (hence restricted ability to escape) and poor record of physical and mental recovery. In this study, the rate of fatal fires from heating devices were higher among older people. The authors noted that smoke detectors, which were often not used, would have been effective in preventing deaths. In a review of older burns victims in Vancouver, the mean age was 76 (65–99) with a mortality of 59% (Hill *et al.*, 2002). The main predisposing factors in this study were alcohol (21.4%), cigarettes (10%) and dementia (33%), with probable dementia in 44%. The authors concluded bathtub burns are strongly associated with dementia and argued for more preventive care, including routine checking of bathtub temperatures.

Not surprisingly, falls are three times more common in hospitals and nursing homes then in the community. On a physical level, falls in these locations are more likely to produce fractures and soft-tissue injuries (Rubinstein and Powers, 1999). Many falls have multiple aetiologies and risk of falls exponentially rises as the number of risk factors increases (American Geriatric Society, 2001). In the hospital setting the rate of falls is reported to be between three and ten falls per month for a 25-bed ward. Pooled data from controlled trials have shown a reduction of 25% in falls following interventions (Oliver *et al.*, 2000). A good example is having beds visible from the nursing station (Vassallo *et al.*, 2000).

Hospital bed rails are prone to produce more severe injuries related to falls (Hanger *et al.*, 1999). Observational studies have found 60 to 70% of all falls occur from the bed or bedside chair. Up to 90% of falls from bed in hospital happen despite bedrails. Moreover, use of bed rails has been known to cause deaths (Oliver, 2002). Deaths from bed rails are preventable and often under-recognised. In one study of 74 such deaths, 70% were from entrapment between the mattress and the bed rail, 18% were from compression of the neck within the rails and 12% were from being trapped by the rails (Parker and Miles, 1997). Importantly, it should be recognised that the use of physical restraints, in whatever form, is an invasion of the personal space of the older person and should be avoided. If done, it should be done with the consent of the patient or the guardian, and to the absolute minimum.

Another major consequence of hospitalisation for older people is delirium. A sudden change of environment in association with medical illness will lead to delirium, which is potentially preventable (Inouye, 1994). If older people can stay in familiar surroundings the chances of delirium can be minimised. In a randomised controlled trial of home versus hospital treatment for medical illness, delirium and incontinence were significantly reduced in patients treated at home (Caplan *et al.*, 1999).

Bridging institution-based care and full independent living, the provision of intermediate or transitional care units is a relatively recent trend. The National Health Service in the UK has committed funding for these. There is some controversy regarding this initiative. While some agree that they are for providing older people with time to recuperate from acute illness, hospitals often regard them as a mode of 'unblocking' hospital beds and saving money. Steiner has said, 'there is a political rhetoric about therapeutic benefits that cannot be matched by the realities of everyday practice – unless providers change their practice with older patients' (Steiner, 2001). Indeed these units, in the UK, when led by nurses, did not save money since the length of stay was longer than in traditional hospital-based care following the acute episode (Griffiths *et al.*, 2001). Another major issue with these units is that they all assess the patient outcome by the same standardised tool, whereas patients' needs are more likely to be heterogeneous. For example, one patient may need to negotiate stairs whereas the other may only have to walk to the toilet three metres away. The advantage of these units are that they give the frail person time to recover, set the goals for going home and start the dialogue with the health care professionals.

It is well known that older people do not get sufficient time or opportunity to weigh up the pros and cons of moving from their homes into residential care. The typical scenario is an older person, living at home with community support, suddenly getting unwell and delirious. The family and hospital may then decide to get the patient into residential care, without giving the patient much time or opportunity to come to terms with this decision. Nevertheless, moving into a residential or nursing home is one of the most traumatic transitions an older person has to make and this is often made when the person is unwell and incapable of deciding. Transitional care units will fill this vacuum.

Another important issue is older people and hoarding. Chronic self-neglect and hoarding is described as Diogenes' syndrome (Clark *et al.*, 1975). When people like this are admitted to hospital, there is often reluctance from the health care professionals to discharge them to their homes. These types of situation warrant competency assessment in a consistent and rigorous manner. It must be a well-researched and thought-out process (Lowe *et al.*, 2000). Lowe *et al.* describe the case of an 82-year-old lady, who was living alone in 'filthy' surroundings. The niece took her away for a 'holiday'. When seen by the junior medical staff of the hospital, who challenged her ability to go home and the state of the house, her response was, 'How would you like it, if a person turned up one day, took you away from your house, said you were going on holidays, and then flew you to somewhere like this, where you can't even get away? You wouldn't like that, would you?' This lady needed more detailed and lengthy assessments, before a decision could be made.

The ability to drive is obviously an important mobility issue for older people which can maintain their quality of life. However, at the same time, accidents caused by older drivers with deteriorated physical and/or mental states are a major public concern. Many older people selectively cope with their decline by compensation (Readt and Ponjaert-Kristoffersen, 2000). They do this by selection, optimisation and compensation. They selectively prepare the trips, choose time and destination, and avoid complex traffic situations. Readt *et al.*, in their study of 84 healthy 65-year-old and older drivers, with a 'Mini Mental Score' of 27 or above, had shown that drivers who compensate in these ways do cause fewer accidents.

Due to both ageing populations and cheaper air travel, more older people are travelling by air. This, however, leads to an increase in associated medical problems. Most commercial aeroplanes travel at altitudes between 28,000 and 45,000 feet. The cabin pressure is usually maintained at equivalent to 6000–8000 feet, where the alveolar oxygen tension is reduced to 71 mm of mercury. The humidity in the aircraft is low and the passengers are often confined to cramped seats. Dehydration is also a major consequence on flights. Older people are more likely to suffer from these challenging conditions because of their multiple co-morbidities and poor physiological reserves. Ischaemic heart disease and worsening congestive cardiac failure are the two common medical emergencies during air travel. Neurological conditions, especially epilepsy, are also common. Most older people have anxiety about flying and most of the formalities at the airports are stressful for them. Missed and misplaced medications are also a common occurrence in travel. As Low and Chan (2002) suggests, before a flight a thorough medical review is preferable.

Another cohort of older people worthy of a mention are located in prisons. However, their health and health care needs are not widely known or studied. In a study of 203 male prisoners in the UK, 85% had more than one illness in the medical record and 83% revealed more than one chronic illness in interview. These subjects certainly pose specific health and space-related issues for prison authorities (Fazel *et al.*, 2001).

In countries like Australia, older people living in rural and remote regions face different problems. They have lower life expectancy, higher mortality, higher hospitalisation rates and lower socio-economic conditions. Access to health and medical services are sub-optimal. For example, indigenous people make up 2% of the overall Australian population and in remote areas they constitute up to 26% of the local community. Geographical location of services will become a major challenge, as the population grows older (Gibson, 2002). These people are less likely to get any of the services their counterparts receive in urban areas.

With the increasing number of frail elderly people living in the community, the traditional role of services is also changing. Some of the day centres, traditionally for social activities, are providing more and more 'care' for their clients. Some of these people would have been institutionalised in the past (Gross and Caiden, 2000). Assisted living facilities are becoming a very popular form of residential care for older people. These facilities provide assistance in activities of daily living (ADLs), meals, laundry and housekeeping. They have organised social activities. This way the resident's care is individualised and the care is needs-based.

On a related note, retirement villages have the concept of flexibility built into them with graduated levels of care. 'Ageing in place', in this guise, is increasingly popular in most countries. However, the ageing-in-place concept works only when residents get graduated care in the same place throughout their stay. This is often restricted by funding and regulatory bodies. In Kansas a study found that this concept was compromised by facility admission and discharge policies (Chaplin and Dobbs-Kepper, 2001). If the concept of ageing in place has to work we need flexible and needs-based programmes where the residential care can be delivered consistently with the declining physical and mental health of the residents, as time goes on.

The concept of retirement communities was originally developed in the USA and now it exists in Australia and Canada. The emphasis is more on lifestyle, entertainment, recreation and fun than on frailty and ageing. Many such facilities are user-friendly for older people, with easy layout and safe design. They have a positive image of ageing and are a new form of age-specific space. However, many poorer people are excluded from these because of market forces.

Residential care is certainly not individualised and is almost always institutionalised. Most residential care places have the following factors in common (Hugman, 2001):

- all aspects of care are in one place,
- most of the residents have no prior relationship,
- staff have power over the residents including daily routine and space,
- there is separation of the group from the rest of the society and
- there is limitation of individuality.

Nursing homes are the main providers of long-term care for older people and a considerable number of these people live their last months there. In 1995, in the UK, they provided care for 158,000 residents. With the demographic transition there is some anxiety about whether all countries should be providing more nursing-home care beds. Moreover, the frailty and acuity of the residents are reportedly increasing too (Turrell, 2001). According to Turrell, markers of good quality of these places are:

- prescribing practice,
- access to medical care,

- access to nursing care,
- access to rehabilitation and
- patient assessment and documentation.

At present there has been little research to compare hospital care to nursing-home care, even though there is potential for these different facilities to provide the same form of care. Older people in nursing homes (and residential homes) experience barriers to relationships with families and friends. Lack of sufficient space makes this almost impossible and often difficult. Families will have to fit in with the facility's space and time. Often private conversations and personal relations are difficult. Many people in residential and nursing home care may wish to share a room or bed. Unfortunately, this is often discouraged because of the rules and regulations and family objections. It is not uncommon to see a husband and wife who have been married for over 50 or 60 years to be separated when one has to go into residential care. When both need to go, they may be even in two facilities.

One of the most challenging behavioural disturbances in dementia is wandering. This can be addressed by environmental manipulation, medications and activity programmes, even though there is no evidence for unequivocal effectiveness for any of these (Lai and Arthur, 2003). It is not known why people do this although one may speculate that they are trying to escape to the familiar space and old surroundings. However, what is well known is when the wandering person is restrained, they become more agitated and aggressive. In practice the best remedy for it is unrestricted space to wander.

Death is a common event in nursing homes. For example, 31% of people will die within 6 months and 43% within 12 months of institutionalisation (Australian Institute of Health and Welfare http://www.aihw.gov.au). However, it is concerning that facilities are not typically equipped to deal with this issue on either a physical or a psychological level (Lloyd, 2000).

In summary, this chapter has introduced, very practically, a wide range of issues related to ageing bodies and space. It has shown that the negotiation of space play a major part in older people's lives. From the macro to the micro scale, health, welfare, community and family are all interrelated around uses of space. The issues discussed, however, indicate how current approaches to accommodation and care are often reactive and locally specific. It is important that health care professionals and policy-makers formulate proactive, consistent and long-term approaches. By investing in safer and more user-friendly spaces, and being aware of older people's needs, we are essentially investing in our own futures.

9

ENVIRONMENT AND PSYCHOLOGICAL RESPONSES TO AGEING

Robyn Findlay and Deirdre McLaughlin

Introduction

Most of the theories on the environment and the psychology of ageing have been built on observations of older people in institutionalised care. However, this cohort is not a representative sample of the older population as only around 5% of people over the age of 65 live in such an environment (Bell *et al.*, 1996), and most are there because their failing mental or physical bodies are no longer able to cope unassisted. Consequently, these theories over-emphasise the decrements of old age (Carp, 1987; Ranzijn, 2002).

During the last decades, there has been a marked demographic change, with people now living longer and in better health than ever before. As a result, the inadequacies of older theories for explaining the ageing process have become apparent and there has been a shift towards a more positive view of ageing that emphasises how well older people adjust to the changing environment (e.g. Ranzijn, 2002). Nevertheless, as noted by Baltes and Smith (2002) there is indeed a time, if one lives long enough, when no matter how hard one tries to be proactive within one's environment, the ageing process makes it very difficult for even the most resilient to succeed. At this point, older people are 'at the limits of their functional capacity and science and social policy are constrained in terms of intervention' (p. 1). Baltes and Smith label this stage the fourth age, which roughly equates to being over 85 years of age and is a time that 'tests the boundaries of human adaptability' (p. 13). Of course, 85 is only an arbitrary age, with the ability of a person to adjust to environmental change being more related to personal attributes than to age. Some people of 85 are better adjusted than others of 65.

In this chapter we examine how older people respond via behaviour, feelings and emotions to changes in their environment, a key mediating factor being the individual attributes of the older person. The focus of this discussion is on transactions between the psychological, social and physical environment in old age, and does not include consideration of biological, economic and political environments. We begin by examining the key theoretical perspectives in environmental psychology on ageing, then consider the person–environment transaction cycle and the role of individual differences in determining adaptive responses. The attention then turns to the repertoire of psychological strategies older people use to deal with changes in their social and physical environment. The following discussion is limited, however, by its focus primarily on Western culture.

Key theories

Ecological theory of ageing

The evolution of one of the most prominent theories in the field, the ecological theory of ageing (ETA; Lawton and Nahemow, 1973), has mirrored the shifts in research focus in the rest of the field. ETA has been influential in the work of researchers and practitioners in a variety of ways over the last 30 years (see Nahemow, 2000 for details). ETA, which has adaptation level as its central concept, was based on the environmental docility hypothesis (Lawton and Simon, 1968). That hypothesis depicted older people as passive players with declining competence who were subject to the demands of the environment (environmental press). According to the environmental docility hypothesis, as personal competence declines the environment becomes a more potent determinant of adaptation level. The shortcomings of this hypothesis were not immediately apparent because it accounted quite well for behavioural responses of the majority of older people in institutionalised care and it was aligned with society's perceptions of older people in the 1960s and 1970s. Gradually, with the help of criticism from the architects of the complementary/congruence model (Carp and Carp, 1984) and the person–environment congruence model (Kahana and Kahana, 1982), the limitations of the environmental docility hypothesis were exposed. Subsequently Lawton (1985) amended the theory to include the complementary hypothesis of environmental proactivity, which acknowledged the reciprocal nature of the person–environment relationship and accounted for the behavioural responses of older people of above average competence for whom the environment was more of a resource than a controller of behaviour.

This amendment to ETA was the last major development in the field for almost a decade. Indeed, the *Handbook of the Psychology of Aging*, which had included a chapter on environment and the psychology of ageing since its first edition, omitted a chapter in its 1996 edition because 'there had not been sufficient progress since the previous handbook edition' (Birren and Schaie, 1996: xvi). The chapter has reappeared in the 2001 edition with an acknowledgement that some progress has been made, albeit disparate, and that greater focus needs now to turn to the different cohorts of older adults, such as the baby boomers, and how they respond to the changing environment (Wahl, 2001).

Although there have been no overarching theories of the psychology of ageing with regard to environmental influences, several theoretical frameworks other than ETA have informed development in the field and are discussed in the following sections.

Selective optimisation with compensation

One of the most notable recent theories is selective optimisation with compensation (SOC; Baltes and Baltes, 1990; Baltes and Smith, 1997). SOC assumes that in old age the losses begin to outweigh the gains and there comes a time when individuals cannot rely on social support to fully compensate for their losses. Then a key strategy for effectively managing the ageing process is to invest more and more personal resources into maintenance and repair rather than growth (Baltes and Smith, 2002). Baltes and Smith (p. 21) provide an excellent example of the theory by describing how the pianist Rubinstein continued to successfully perform concerts into his 80s:

> He played fewer pieces, but practised them more often, and he used contrasts in tempo to simulate faster playing than he in the meantime could muster. Rubinstein reduced his repertoire (i.e. selection). This allowed him the opportunity to practise each piece more (i.e. optimisation). And finally, he used contrasts in speed to hide his loss in mechanical finger speed, a case of compensation.

Disengagement theory

Disengagement theory (Cumming and Henry, 1961) posits that older people withdraw from society as a normal process of ageing and that society ceases to depend on them. According to Cumming and Henry, this withdrawal is a healthy way of coping in old age. However, objections to their theory are numerous and are summarised adequately in Youmans (1969). Indeed, Havinghurst (1963) built an opposing activity theory based on the same data as used by Cumming and Henry. Activity theory proposed that older people combat society's withdrawal by continuing their roles or substituting them with new ones.

Control-related theories

One of the most important personal resources in the environment–ageing interaction is a sense of control over the environment (Cerrato and Fernandez de Troconiz, 1998; Chowdary, 1990; Rodin and Langer, 1977). It affects how losses are perceived and subsequently compensated for. Loss of control is related to feelings of helplessness and can be detrimental to psychological functioning. The reason why perceived control is so important in old age is not yet clear, but it may be that it reduces the feelings of threat associated with difficult environments (Pearlin and Skaff, 1995). Another explanation is that it might predispose one to act and mobilise social support when it is needed (Brandtstadter and Renner, 1990). Schulz and Heckhausen (1999) suggest that striving for control has an evolutionary basis and is advantageous to all species that have the potential to influence their environment.

Several theories have been formulated around the issue of ageing, control and the environment. There are lifespan theories that differentiate between primary and secondary control (Heckhausen and Schulz, 1995), theories that emphasise adaptive responses (Brandtstadter and Renner, 1990; Folkman and Lazarus, 1990; Kahana and Kahana, 1983), and socio-emotional selectivity theory that seeks to explain the reduction in social networks that is apparent in old age (Baltes and Carstensen, 1999; Carstensen, 1991; Carstensen *et al.*, 1999).

Life-course theories

Life-course theories (e.g. Erikson, 1959/1980; Neugarten, 1975) depict old age as one stage in the developmental process. For example, Erikson's theory of psycho-social stages in human development depicts old age as a stage of ego integrity versus despair when older people reflect on their life either with a sense of satisfaction that they have lived it well, or with a sense of despair over their losses. According to the theory, it is also a time of generativity (productivity and creation versus stagnation). In 1986 Erikson expanded this concept to encompass what he termed 'grand-generativity', which will be discussed later in this chapter.

Adaptive responses and the person–environment transaction cycle

Contrary to ageist stereotypes, older people are not a homogeneous sector of society. On the whole, they are very resilient individuals who use a variety of personal resources to maintain their well-being and satisfaction amidst daily environmental challenges (Baltes and Baltes, 1990; Cerrato and Fernandez de Troconiz, 1998). Each older person has a unique psychological environment that includes different personal attributes and life experiences, as well as a physical and social environment that impacts differentially on them. All of these factors, plus physical well-being, economic status and the political climate, determine how they perceive and respond to environmental challenges (Cerrato and Fernandez de Troconiz, 1998; Daneffer and Uhlenberg, 1999; Rubinstein, 1989). Most people, including older people, pay little attention to their environment until it poses some threat or challenge to them. At this point they respond, either by taking action to directly change the situation, or by using emotion-based strategies to indirectly change the situation (Lazarus and Folkman, 1984). The response that they choose adds to their experience in dealing with the environment and may in turn change their perception of themselves and their environment for future encounters and so on as shown in Figure 9.1. The focus of this chapter is on the transactions between the psychological, social and physical environments (see also chapters 11 and 14).

Psychological environment

Individual attributes

The psychological environment includes personality, perceptions of self (e.g. self-concept, self-esteem, self-efficacy), perceived control and privacy preferences. Evidence for the stability of these psychological attributes across the lifespan is largely inconclusive.

There is a large body of research that suggests that personality traits show high intra-individual stability over the adult lifespan (e.g. Costa and McCrae, 1988; 1994; Schaie, 1996; Schaie and Willis, 1991), but other studies indicate there are age-related differences. For example, for the 'big five' personality factors (Goldberg, 1993; John, 1990; McCrae and Costa, 1990), Costa and McCrae (1992) demonstrated that, with ageing, there are small but consistent declines in neuroticism, extraversion and openness to experience, and small increases in agreeableness and conscientiousness. These findings are supported by cross-cultural data showing the same patterns of age differences in samples from Croatia, Germany, Italy, Korea and Portugal (McCrae *et al.*, 1999).

This pattern of age-related changes in personality is compatible with studies that have examined age-related differences in adaptive responses and affect regulation. The latter literature consistently indicates that older people are more likely to regulate their behaviour in response to an environmental challenge than take direct action to change the characteristics of their environment. They use strategies such as accommodation, suppression of emotions, reinterpretation of situations and avoidance of conflict by changing attitude and by acceptance of negative events to accomplish this (Blanchard-Fields *et al.*, 1995; Brandstadter and Greve, 1994; Castro *et al.*, 1995; Diehl *et al.*, 1996; Wister, 1989). Interestingly, Labouvie-Vief *et al.* (2000), in a comparative study of adults from the United States of America and China, found that the age-related differences in personality are more pronounced in the Chinese; that is, in cultures where individuals have been socialised into a stricter code of emotional regulation.

Studies such as the Berlin Ageing Study have found that in the fourth age there is a distinct decline in psychological functioning (Smith, 2001), with changes in intellectual status being most marked. The decline in cognitive functioning has been attributed to environmental influences such as a lack of stimuli or reinforcement from the environment (Labouvie-Vief *et al.*, 1974), role loss as a result of retirement or children leaving the family home (Kuypers and Bengston, 1973), and ageist stereotypes (Kuypers and Bengston, 1973; Mitsch Bush and Simmons, 1981).

Rodin (1990) and Weisz (1983) suggest that the environmental events

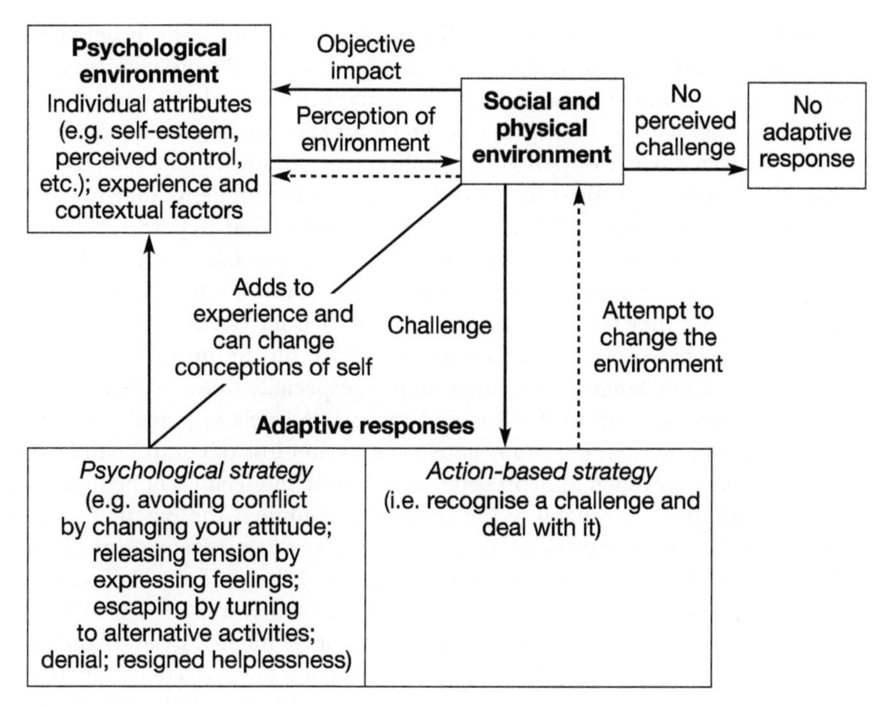

Figure 9.1 Adaptive responses and the person–environment transaction cycle

accompanying ageing can limit the range of outcomes that are actually attainable. This limitation can set up a downward spiral by lowering self-esteem and self-efficacy and can subsequently diminish a person's sense of control over the environment. The perceived loss of control, in turn, can result in more negative responses to environmental challenges. More negative responses then feed back to lower concepts of self even further, and so on. Whether this sequence of events unfolds for an older person depends primarily on how they interpret and respond to the challenges over the lifespan.

At this point, though, some clarification of terms such as self-concept, self-esteem, self-efficacy and perceived control is necessary. Self-concept is one's knowledge, feelings and ideas about oneself; that is, the overall picture one has of oneself (Carlson and Buskist, 1997). Self-esteem is the personal attitude that individuals maintain regarding their own worth and social value (Rosenberg, 1965; Twigger-Ross and Uzzell, 1996), and is an important coping resource (Liebkind, 1992; Thoits, 1995) that can affect the way they perceive their environment and their control over it. Low self-esteem is associated with lack of motivation and confidence, and with limited skill to change a challenging situation (Lund *et al.*, 1993), whereas high self-esteem, which is associated with higher self-efficacy, is a buffer against environmental challenge and can empower a person to respond more proactively (Thoits, 1995).

Self-efficacy is an individual's belief in their ability to perform behaviours required for attaining a specific goal (Bandura, 1989). People with high self-efficacy have confidence in their ability to perform the behaviours needed to overcome obstacles and satisfy needs. The most effective mechanism for boosting self-efficacy is performance accomplishment (Smith, 1993). Therefore, as Bandura emphasises, there is a need to create environments that maximise opportunities for accomplishment; an environment that is perceived as unmanageable constitutes a threat to self-efficacy (Twigger-Ross and Uzzell, 1996).

Self-efficacy is strongly related to perceived control, which is the sense a person has that they regulate what happens in their life (Lachman and Burack, 1993). Perceived control has a critical influence on the meaning that older people assign to changes in the environment, especially those related to losses, and how they respond to the losses. Appropriate levels of perceived control result in a positive response to negative events, but loss of control can lead to negative outcomes such as depression, withdrawal and feelings of helplessness.

Research on controllability has generally examined the extent to which individuals attribute outcomes to internal sources (contingent on their own behaviour or characteristics) or external sources (under the control of others or luck, chance or fate); that is, their locus of control (Rotter, 1954; 1990). In general, internality is viewed as the more desirable of the two styles (Blanchard-Fields and Chen, 1996) and is associated with higher levels of education, better health, greater wealth and youth (Heckhausen and Baltes, 1991; Heckhausen and Schulz, 1995). Although internality may decline with age, the fact that older people do not display the negative emotions or distress commensurate with the loss (Blanchard-Fields and Chen, 1996) may be due to their use of secondary control which allows them to accept age-related losses more easily.

Primary and secondary control are two distinctions made by Heckhausen and Schulz (1995). Primary control is the ability to maintain influence over the external environment via direct, instrumental action. Secondary control involves acceptance of realities that cannot be altered, and exertion of control over the psychological consequences of the challenge by changing the internal world of the individual. Secondary control allows a person to minimise and manage losses in primary control. Schulz and Heckhausen (1999) suggest that the preference for primary control is a fundamental human universal, but this position has been challenged by Gould (1999), who argued that some Eastern cultures value secondary control more highly.

There is evidence that perception of control is a particularly important resource for older people to preserve their psychological well-being (Chowdary, 1990; Langer and Rodin, 1976; Rodin, 1986b; Rodin and Langer, 1977). A study by Langer and Rodin (1976) of 65- to 90-year-old residents in an institutionalised care setting showed that many of the characteristics usually attributed to the ageing process such as depression, helplessness and physical decline are related at least in part to loss of perceived control. When residents in Langer and Rodin's study (1976) were encouraged to make decisions for themselves they became happier and more active. Similar results were found in a study by

Schulz (1976) of older people (67 to 96 years old) living in institutionalised care. A follow-up of the Schulz study concluded that it is important that the perceived control is generated through the older person's own efforts rather than induced by another figure on whom the older person may become dependent (Schulz and Hanusa, 1978).

Experience and contextual factors

According to lifespan theory, the range of personal resources available to respond to environmental challenges in old age is related to previous life experiences (Carstensen *et al.*, 1995; Daneffer and Uhlenberg, 1999). That is, differences in both impact of and response to the environment depend on choices one has made throughout life (e.g. whether they have built and maintained a social network, whether they have acquired specialised knowledge and skills, whether they have remained physically active etc.). In old age, these life-course choices expand or limit the resources available to enable the individual to cope with their changing environment (Cerrato and Fernandez de Troconiz, 1998: 30). These choices, as well as previous social and cultural experiences, shape the goals and demands that a person brings to later life transactions and influences their appraisal and interpretation of a situation and their motivation to respond in particular ways to environmental challenges (Blanchard-Fields and Chen, 1996).

Adaptive responses

There have been several categorisations of adaptive responses, but with the notable exception of 'selective optimisation with compensation' (Baltes and Baltes, 1990), most can be grouped into the two broad categories offered by Lazarus and Folkman (1984). That is, as either problem/action-focused (altering the element of the environment that is causing distress or anxiety) or emotion-based (adopting psychological strategies to make the situation more beneficial to themselves). These distinctions are also consistent with the environmental docility–proactivity hypotheses of Lawton (1985), and the five coping styles outlined by Kahana and Kahana (1983). The latter can be collapsed to problem-focused (recognising a problem and dealing with it) or emotion-based (avoiding conflict by changing your attitude, or releasing tension by expressing feelings, or escaping by turning to alternative activities, or resigned helplessness).

In old age there is a move away from problem-focused responses to emotion-focused responses (Cerrato and Fernandez de Troconiz, 1998; Folkman and Lazarus, 1990). Adaptation to the environment is 'increasingly achieved through changes in perceptions of the self in relationship to the environment' (Coleman, 1993b: 86). When confronted with environmental challenges, older people would rather engage in psychological processes of adaptation such as restructuring the meaning of the situation, suppressing emotions or trying to manage the effects of the situation, than take direct action to change the characteristics of their environment (Blanchard-Fields *et al.*, 1995; Castro *et al.*, 1995; Wister, 1989). The results of the Berlin Ageing Study showed that older people use these strategies to shape or redefine the meaning and importance of environmental challenges in order to maintain their sense of well-being, life satisfaction and sense of control (Baltes and Mayer, 1999). Baltes and Smith (2002) suggest that it is one of the 'best insurance policies for well-being in old age' (p. 10).

Similarly, Brandtstadter and Renner (1990) found that older people tend to accommodate (change their goals in relation to obstacles) rather than assimilate (actively change their environment to meet their ongoing goals). This shift in focus is accompanied by a preference for the simple and familiar. According to Lawton (1990), familiarity is a major environmental need of older people, and underscores the reason why most people prefer to remain in their own homes rather than relocate. It is a means to sustain their self-esteem. Newton *et al.* (1986) suggest that the narrowing of focus to the self and the simple and familiar is an adaptive response to ageing that 'ensures survival by utilising remaining energy to meet one's essential needs'. Whether these same coping preferences will be the modus operandi of the baby boomer generation remains to be seen.

The following two sections will outline some specific examples of how the social and physical environments can impact on older people and how older people in turn adapt to these challenges.

Social environment

The importance of the social environment

There is an interaction between the social and psychological environments such that a supportive social environment can enhance an older person's psychological resources (Bondevik and Skogstad, 1998; Bosworth and Schaie, 1997; Fratiglioni *et al.*, 2000; Lang and Baltes, 1997; Lang and Carstensen, 1994). At the same time, the older person's resources can facilitate the development and enhancement of social support (Dunkle *et al.*, 1994). Conversely, unsupportive social environments and a lack of personal resources can have the opposite effects.

Erikson, in his theory of grand-generativity (Erikson *et al.*, 1986), suggests that a person's interaction with the environment, and particularly the social environment, is important for giving purpose to later life. According to Erikson, grand-generativity activities such as assisting friends and neighbours, caring for the wider community, volunteering, and expressing concern for environmental issues allows older people to be socially proactive and valued, and to leave a permanent legacy of themselves (Keyes and Ryff, 1998; McAdams and de St Aubins, 1992).

Ageing and the declining social network

There is robust evidence that the size of one's social network gradually declines in old age and much of this decline can be attributed to loss that is often beyond the person's control (Lang and Carstensen, 1998; Okun and Keith, 1998; Pinquart and Sorensen, 2001). Loss can take many forms, such as loss of home and known environment after moving to institutionalised care, loss of social roles especially after retirement, loss of mobility through frailty, loss of spouse and peers through death, loss of driving ability and loss of self-esteem often through the effect of ageist stereotypes. Although older people are generally more resilient than younger people in adapting to losses (Rodeheaver and Datan, 1988), loss can diminish the opportunities to make and maintain social contacts. Nevertheless, it is what the diminished social environment means to the older person that counts, and not necessarily how it appears to an outsider.

On the one hand, the older person may perceive the diminished social environment as an indication that they have little worth or social value. If they do not have the resources to respond effectively, they may suffer adverse consequences such as loneliness, social isolation, depression and suicide (Findlay, 2003; Prince *et al.*, 1997). The impact of reduced social networks in old age has been demonstrated by Pinquart and Sorensen (2001), who found that that between 5 and 15% of people over 65 report that they frequently feel lonely and an additional 20 to 40% occasionally feel lonely (e.g. Prince *et al.*, 1997). For the oldest-old, the changed social environment often impacts more sharply and studies have shown that up to 50% of people over 80 years of age report that they often feel lonely (Pinquart and Sorensen, 2001; Smith and Baltes, 1996).

On the other hand, socio-emotional selectivity theory (Carstensen, 1991; Carstensen *et al.*, 1999) suggests that the reduction in social networks and social participation should be seen as a proactive response by the older person to manage their social world. According to the theory, older people do not simply react to the social environment. They regulate their social network depending on whether they perceive their future time as expansive or limited. When they see their time as limited, emotionally meaningful goals are pursued and the quality rather than the magnitude of their relationships with others assumes greater importance. They reduce their social network to conserve energy, and select those people for continuing relationships who bring about positive emotions and strengthen their self-esteem. Close and long-term contacts who fulfil their socio-emotional needs are more likely to be maintained than those who focus on providing information (Carstensen *et al.*, 1996; Nahemow, 2000; Pinquart and Sorensen, 2001). In terms of Baltes' theory, the reduction of one's social network is a function of selective optimisation. The extent to which relationship regulation depends on the older person's psychological resources (or action potentials) is, however, not yet well understood (Lang *et al.*, 2001).

Ageist stereotypes

The social environment can impact on older people via ageist stereotypes. According to Kuypers and Bengston (1973), older people can be induced into social and cognitive incompetence by their losses especially if they start to adopt society's expectations of their behaviour. If they become receptive to ageist attitudes they may come to see themselves as incompetent and begin to set up a downward spiral where 'failure in one aspect of daily life is used to justify and excuse failure in other areas' (Coleman, 1993a: 80). A study by Levy and Langer (1994) emphasised the potential influence of ageist stereotypes. Levy and Langer found that older adults from cultures such as China, where older people are revered, performed better on memory tests than those from cultures (e.g. the USA) where older people are less revered, suggesting that cultural stereotypes can become self-fulfilling prophesies that impact on memory loss in later life. The role of ageist stereotypes in cognitive decline is an area that needs further investigation.

Physical environment

There has been very limited prior research on the links between the physical environment and the psychological adaptation of older people to either continuity or change within their physical environment. As people age, their interactions with and reactions to their environment may alter to account for changes within their physical and psychological capacity. Possibly, the most salient issue for older people and their relationship with the physical environment pertains to where they live. To continue to reside within their own home, many older people may require a support structure that is primarily community-provided, although in many cases older people are maintained within their own homes by their family. However, this may not be appropriate for all older people and for some the only choice may be relocation to a supported care environment. Another important component of the interaction between older adults and their physical environment involves their ability to drive a car and thus have access to those areas of the environment which are not immediately proximate to their residence. Changes in any of these components of the physical environment, substantial or otherwise, may challenge the psychological well-being of the older person. The way they respond to these changes will be examined in further detail.

Relocation into aged-care residences

Prior research conducted on older persons residing within institutions has shown that they suffer from more negative self-evaluations, lower self-esteem and fewer intimate relationships and demonstrate less interest in interdependence and interactions with others than older people still living within their own homes (Antonelli *et al.*, 2000). Furthermore, institutional care may even hasten the course of an illness. It has been suggested that these negative outcomes may be fuelled by the impersonal and structured nature of the institutional environment (Antonelli *et al.*, 2000). Because of the long-established behaviour patterns demonstrated by older adults, they may find it difficult to adjust to institutionalised care. However, this is not invariably a product of institutionalisation, but rather maladaptive strategies utilised when an older person is placed in a physical environment which they may perceive themselves as having little control over and little ability to change. According to Bell *et al.* (1996) the response to relocation may be predicated on the degree of control over the process that the older person feels that they are exerting. Perceptions of lack of control result in feelings of disempowerment and may eventually elicit a learned helplessness response. For example, Baltes *et al.* (1992) demonstrated that residents in nursing homes have dependent behaviour consistently rewarded by staff while competent, self-motivated behaviour is ignored or actively discouraged. Therefore residents may develop doubts about their own ability to adequately perform tasks for themselves and may become reliant on staff.

The factors which underpin an older person's relocation to an institutional setting may also impact on their ability to utilise positive adaptive strategies in adjusting to their new environment. For example, people frequently enter residential care settings because of illness, death of spouse or carer or lack of ability to attend to their personal care within their own home. All of these issues can lead to perceptions of loss of control, particularly where the decision to enter care has been mainly taken by the family. The psychological strategies utilised by older persons who are relocated from an own-home environment into an institutionalised environment will depend on the individual attributes which the individual can bring to bear on the situation. The importance of a positive sense of self-esteem has been mooted as a critical factor in adjustment to institutional care. Antonelli *et al.* (2000) demonstrated that institutionalised older people described themselves more negatively than older people still living in their own homes. Further, older people living at home had higher levels of self-esteem than those living in care. It may be that negative self-concept and low self-esteem are antecedents to the move to institutional care, particularly where the underlying rationale for the move has been the inability of the older person to adequately care for themselves. Conversely, lack of control over the environment and the fostering of dependency may of themselves be sufficient to stimulate maladaptive coping strategies within older people. Therefore the response provided by the individual can actually help reframe the environment (Lawton, 1990). That is, the way in which older people respond in an institutional setting can alter the way in which the institution treats them. Other factors which may impact on the adjustment of older people to relocation is the type of accommodation into which they are placed; that is, shared or single units, size of development, type of institution (e.g. nursing home, hostel or single living units) and proximity to facilities. However, key determinants of successful adaptation to relocation appear to be a sense of control over the process and positive self-esteem, which allows for the utilisation of adaptive strategies to maximise the person–environment fit.

Maintenance within a familiar home environment

For many of the elderly, particularly those who are mobility impaired, remaining in the home in which they have lived for years fulfils a major environmental need – that of familiarity (Wahl, 2001). Further, attachment to place has been demonstrated to be higher among older persons (Hidalgo and Hernandez, 2001) for whom the home may provide a link between their past, more active selves and their current physical and psychological status. The home may be perceived to be a symbol of personal achievement, closely allied with the older person's self-concept, which older adults consistently frame in terms of their activities, attitudes and life circumstances, including their possessions (Antonelli *et al.*, 2000; Lawton, 1990). Research on attachment to home supports the view that the strongest link between older persons and their physical environment is the one to their own home (Wahl *et al.*, 1999) and that most would prefer to cope with the challenges of ageing within the framework of their home. Comparisons between institutionalised and non-institutionalised older persons reveal that those remaining in their own homes have more intimate relationships and more positive self-evaluations, are more interested in interactions with others, and suffer less from depression (Antonelli *et al.*, 2000). However, fit, or lack of fit, between personal capabilities and environmental conditions and demands may be the prime factor in determining if an older individual can successfully maintain themselves within their own home.

Wahl *et al.* (1999) have suggested that older people may utilise a wide variety of compensatory mechanisms to increase congruence between their competence and the environment. These mechanisms may be seen as utilising either primary or secondary control over the environment and oneself. For example, physical modifications to the home to allow for decreased physical agility and mobility are an adaptive response involving primary control; while the application of secondary control might involve the acceptance of the limitations imposed by physical mobility and at the same time controlling such possible psychological sequelae as depression.

In short, for many ageing adults, the desire to remain in a familiar environment is strong. The capacity to remain within an own home environment appears to depend on the ongoing ability of older adults to adjust, both physically and psychologically, to the limitations imposed by the ageing process and the congruence of their person–environment fit.

Driving cessation

Driving a car is a tangible and important way in which older people move around their physical environment, yet the psychological repercussions of driving cessation have received little research attention. Many older adults must decide whether to continue or to stop driving as their health, physical functioning or financial resources decline, and this decision has important implications for their ability to maintain social contact and utilise resources within their environment, such as attending to their health care needs and shopping. Yassuda *et al.* (1997) have suggested that older adults derive a sense of identity and independence from driving whereas the loss of the ability to drive implies social disability and fosters dependence on others. Research has tended to support the expectation that driving limitations have negative consequences for older people's quality of life (Fonda *et al.*, 2001). In particular, inability to drive has been associated with a significant decline in out-of-home activity levels (Marottoli *et al.*, 2000). A growing body of evidence (Marottoli *et al.*, 1997; Fonda *et al.*, 2001) has indicated that driving cessation is followed by a significant increase in depressive symptoms, even controlling for factors commonly associated with depression in late adulthood, such as declining health. However, changes in driving patterns need not be associated with negative symptoms if older adults select adaptive behaviour which compensates for loss of, or limitations to, driving ability.

Many older people are able to identify decrements in their physical functions as they age and make adjustments in their driving behaviours to compensate for these, resulting in a reduction of driving, rather than cessation – a modification which can also be described as adaptive behaviour associated with primary control. For instance, worsening visual acuity may lead an older driver to avoid driving at night or in rainy weather while some older drivers prefer to avoid driving on highways and may choose to drive for shorter distances to reduce the impact of fatigue (Fonda *et al.*, 2001). Older drivers who reduce their driving also reduce their chances of developing depressive symptomatology (Fonda *et al.*, 2001) compared with older drivers who cease driving completely. It has been suggested that this increase in depressive symptoms among ex-drivers may be related to the social isolation engendered by transport limitations as well as the health or ageing issues that may have precipitated driving cessation (Marottoli *et al.*, 1997).

Older adults who relinquish driving may choose to adopt a number of strategies to adjust to their situation; for example, some former drivers develop an informal network of ride providers such as family, friends and neighbours. The provision of transport for the older person can have a beneficial effect because of the ensuing opportunity for positive social interactions (Glasgow and Blakely, 2000). Similarly, many older adults, even those who have reduced but not ceased their driving, have described utilising senior citizen's centre or special excursion buses to embark on long-distance trips or evening outings to enhance their social interactions (Glasgow and Blakely, 2000). These approaches to the management of driving cessation or reduction effectively limit the impact of social isolation and may therefore be protective in some measure against depression.

Driving cessation may have a profound impact on the lives of older adults. However, retrospective evaluations of the event have in many instances not been as negative as anticipated, particularly where positive strategies have been adopted to minimise its effects (Glasgow and Blakely, 2000).

Summary

This chapter has examined psychological responses to ageing in relation to challenges in the social and physical environment. Key theories in the field were presented and a person–environment transaction cycle was developed to conceptualise the reciprocal nature of the person–environment relationship and the role of adaptive responses in that relationship. The discussion has demonstrated that as people age they tend to use psychological strategies, rather than action-based strategies, to deal with the challenges posed by their environment, and in doing so the maintenance of a sense of control over the environment is of primary importance to them. These issues are considered further in chapters 6, 7, 8, 10, 11 and 14.

10

AGEING IN RURAL COMMUNITIES

Vulnerable people in vulnerable places

Alun E. Joseph and Denise Cloutier-Fisher

Introduction

The broad demographic trajectory of rural areas in developed economies such as Canada was well established by the early 1980s, with the balance over time of in- and out-migration identified as the chief determinant of regional and local population structures (Bryant and Joseph, 2001). The major structural and geographical characteristics of population ageing were also clear (Rosenberg and Moore, 2001), with advanced population ageing being seen increasingly as a defining characteristic of nucleated communities in Canada's rural hinterland (Joseph and Fuller, 1991). Indeed, in a subset of rural towns, by the early 1980s those aged 65 or older constituted over 20% of all residents, and in some over 30% (Hodge, 1991; Joseph and Martin-Matthews, 1993; Everitt, 1994). Partly related to out-migration of younger cohorts (Fellegi, 1996), such 'top-heavy' community demographic structures also invariably reflected the cumulative impact of in-migration from the local countryside of rural (and often farm) people seeking housing and services (Joseph and Hollett, 1992) or inter-regional (and often urban) migrants seeking amenity and similar housing and services as local migrants (Dahms, 1996; Everitt and Gfellner, 1996). Thus elderly residents of rural towns, whether 'oldtimers' or 'newcomers' in the rural milieu, often found themselves 'growing old in ageing communities' (Joseph and Martin-Matthews, 1993).

Commentaries on rural population ageing have invariably noted implications for the provision of targeted housing and services for the elderly (Joseph and Fuller, 1991) and for the ability of older people to maintain their independence and live out their lives in places of their own choosing (Joseph and Cloutier, 1991). Such commentaries have tended to assume a degree of stability over time in the economic and social fabric of rural settlements and in the services they provide to their elderly and non-elderly residents. However, over the last 20 years or so this assumption has been challenged repeatedly as successive waves of restructuring have reconfigured rural settlement and service systems and redefined community contexts for ageing (Joseph and Chalmers, 1998; Cloutier-Fisher and Joseph, 2000).

In this chapter we examine the experience of growing older in restructured rural community settings. To develop our arguments, we draw primarily upon research from Canada and New Zealand, two countries in which the progressive and cumulative impact of long and short cycles of change have produced considerable diversity in rural community structures (Chalmers and Joseph, 1997; Bryant and Joseph, 2001). Our concern is not with the broad parameters of population ageing, or even of rural ageing, in the two countries, for these have been well established (for example, see Koopman-Boyden, 1993; Joseph and Martin-Matthews, 1993; and Keating *et al.*, 2001). Instead, we seek to draw attention to the 'situated meaning' (Kearns and Joseph, 1997) of rural change as it relates to the experience of growing old in rural communities. Specifically, we deploy a dual focus on people and places as a means of interpreting the critical dynamic between the service needs of elderly people and the servicing capacity of their communities, and identify 'vulnerability' as a social construct that captures the overall tenor of the impacts of rural change on both people and places. While acknowledging that increasing age brings with it challenges to health that transcend place (Joseph and Martin-Matthews, 1993), as elaborated later, we wish to avoid images of individual deficiency that may be associated with the term 'vulnerability'. Rather, we focus on the general threads of vulnerability that challenge successful ageing in rural communities.

Temporally, our emphasis is on the 1980s and 1990s, decades that many regard as being characterised by an extraordinary level and tempo of change in the rural sector (Cloke, 1996; Ilbery, 1998). While some changes during these decades were intrinsic to the rural sector (and often focused on agriculture), most were externally generated and associated with restructuring (Joseph *et al.*, 2001). In Canada, New Zealand and elsewhere, emergent ideologies of fiscal conservatism promoted 'the sanctification of the market as arbiter of all value' (Douglas, 1999: 45) and provided a platform for government downsizing and the centralisation, rationalisation and privatisation of services as a means of spurring flagging economies and reducing deficits (Laws, 1989; Kelsey, 1997; Rice and Prince, 2001). As we shall note later, while not directly aimed at the rural sector, restructuring had special implications for that sector because of a long-established reliance on public-sector investment in support of primary industry and rural services (Joseph *et al.*, 2001).

Throughout this discussion, we define elderly people as those aged 65 or older. In terms of the definition of rural population and places, while sympathetic to the Canadian trend toward labelling all settlements of fewer than 10,000 as rural (du Plessis *et al.*, 2001), we avoid specific engagement with the task of sorting through categorisations of settlements based upon their social system, economic reliance on primary production, distance from large urban centres, and so on (Troughton, 1999; Zapf, 2001) by focusing on specific places that construct themselves as rural. We see this focus on specific places, albeit as representative of the broader experience, as a means of acknowledging complexity in rural communities (Mormont, 1990; Ramp, 1999) that is consistent with our search for the situated meaning of restructuring.

Like Kearns and Moon (2002), we promote a view of place as an operational, living construct rather than as a passive container in which actions occur and histories are recorded. Rural places are seen as dynamic and multi-layered communities, as having their own general physical, economic and political attributes as well as being further characterised by the local institutions, social practices, social structures and human agents that influence daily lives (Laws and Radford, 1998). In recounting experiences of ageing in rural places, we give voice to the opinions of elderly people and their family members, critical viewpoints which are often missing within the research literature (Averill, 2002; Parr, 2003). Nevertheless, as Murdoch and Pratt (1993) caution, giving voice to the stories and experiences of 'others' does not guarantee that we will uncover the relations that lead to marginalisation or neglect, or in this case vulnerability. Indeed, we agree with Parr's (2003) assertion that to understand vulnerability more fully, we must merge our insights about personal experiences with a deeper examination of the spaces and places in which they occur.

The remainder of the chapter is organised in four major sections, the first of which considers the general features of rural populations and rural communities that promote vulnerability. The next section deals with the remaking of rural places as contexts in which people grow old. This discussion draws on a range of theoretical and empirical literature to identify the overlapping effects of short cycles of change associated with economic and social restructuring and longer-term trends in settlement and servicing systems. In both these sections, a distinction is made between the maintenance of everyday living and the ability of elderly people to stay on in their communities year after year. In the next major section, we draw on New Zealand research to illustrate the broad impact of change in rural service systems on the day-to-day and longer-term sustainability of lives in place, and note the pervasive role of technology both as a creator of vulnerability and as a potential means of coping with changes in service systems. This is followed by a complementary discussion, drawing primarily on Canadian research, of the impact of health-sector restructuring on the ability of frail and disabled elderly people to remain in their own homes. In a concluding section, we revisit the notion of vulnerability and consider the transferability of our arguments to other national contexts and other times.

Vulnerable populations and vulnerable communities

The literatures from social work and nursing contribute substantively to our understanding of vulnerability as a social construct and of the nature of vulnerable populations. Broadly speaking, vulnerability is closely associated with notions of disadvantage. Disadvantaged populations that are consistently listed as vulnerable include elderly persons, children, single women with children, members of racial and ethnic minorities, immigrants, lesbians and gay men, the homeless, persons living in poverty, persons affected by mental and physical disabilities, persons with mental illness or those suffering from devastating illnesses such as AIDS, ALS or cancers (Gitterman, 1991; Flaskerud and Winslow, 1998). Inevitably, many of these qualities of vulnerability co-exist. As a social construct, vulnerability also flows from life circumstances and events (such as floods and other natural disasters), from genetic predispositions (such as to breast cancer or sickle-cell anaemia), from lifestyle habits (such as an unhealthy diet, smoking or lack of exercise), and from conditions in the local physical and built environments (such as poor air quality, substandard housing or high incidence of crime).

Older persons stand out among vulnerable populations due to the normative expectation that the overall likelihood of ill health and the associated need for specific forms of health care in populations will increase with age (Evans, 1993). Although most elderly people (over 80%) live independently in the community and draw minimally on health and social support resources (Joseph and Martin-Matthews, 1993), some older persons experience multiple chronic health conditions and have limited personal resources for coping with them. While research suggests that elderly persons living in rural and urban areas have equivalent health status (Reimer *et al.*, 1992), some commentators have noted that rural elderly persons often try to remain independent for longer periods of time and may therefore be less healthy when they finally engage with the formal health care system (Keating *et al.*, 2001). Low income, impaired mobility (both transport and personal), lack of social supports, multiple chronic health conditions and relative isolation from service centres are among the characteristics that contribute to high levels of personal and household vulnerability among the rural elderly (Joseph and Cloutier, 1991; Cloutier-Fisher and Joseph, 2000). While communities can be thought of in terms of the aggregation of their vulnerable residents, there are of course distinct and complementary attributes of vulnerability that exist at the level of settlements.

Rural communities have historically been under-serviced, in comparison both with national standards and (especially) with levels of availability in urban communities. Some of the service deficiencies in rural areas include options for transport to service centres, mental health services, palliative care and respite care (Hodge, 1993; Joseph and Martin-Matthews, 1993). Generally, the likelihood of supportive services such as hospitals, physicians, churches, mental health clinics, buses and mobility vans being locally available diminishes with size of community (Hodge, 1993; Troughton, 1999). For many small rural communities, beyond obvious implications for local availability the trend to centralise service provision into larger centres has also meant a loss of local autonomy and increased feelings of powerlessness relative to large sub-national, national and now global bureaucracies (Cheers, 2001; Zapf, 2001). At the same time, many rural areas have been plagued by increasing levels of poverty and loss of working-age population associated with large-scale and persistent job losses and underlain by restructuring and globalisation trends (Reimer *et al.*, 1992; Douglas, 1999; Troughton, 1999). These multiple challenges to the resilience of communities have dovetailed with social trends such as higher rates of family breakdown through divorce and increasing geographic mobility among family members to problematise informal and family-based support in rural communities (Bagilhole, 1996; Joseph and Hallman, 1998), and this must now be set alongside long-standing deficiencies in formal support services.

Thus the different and distinct vulnerabilities associated with ageing and with living in rural communities create a kind of 'double jeopardy' for elderly people living in rural places. We now consider how rural places were substantively remade in the 1980s and 1990s as contexts for ageing, and we do this as a prelude to examining the changing cost of being in double jeopardy as a vulnerable person in a vulnerable community.

The remaking of rural places: new contexts for ageing

Service provision is a critical link between rural settlements and rural people; services support people and people support services (Joseph, 2002). In considering the implications of the evolving relationship between rural population and services for ageing people and their communities, we turn now to the distinction between the short-term impact of economic and social restructuring and the long-term effect of a complex set of changes embedded in shifting technologies, demographic structures and lifestyles (Joseph, 1999; Joseph *et al.*, 2001). While analytically convenient, this distinction is somewhat artificial, of course, in that restructuring has involved changes in attitudes, beliefs and behaviours (which are important components of 'lifestyle') and the application of new ways of doing things (a manifestation of the use of 'technology'). This said, it remains important to assess the synchronicity of short and long cycles of change and to gauge the magnitude of their implications for rural communities and their elderly residents.

Long cycles of change

The pattern of settlement in rural areas has evolved over time; it is the cumulative record of past equilibria between population and servicing systems. The arena society model, first developed in Scandinavia (Persson, 1992), and subsequently generalised by Fuller (1994; 1997), presents transport technology as a key transformative agent in the three-stage evolution of (rural) settlement systems over time; from the *short-distance*, through the *industrial*, to the *open* rural society. In the short-distance (pre-industrial) society, reliance on horse-drawn transport resulted in a settlement system in which the economic and social spheres of rural life overlapped almost perfectly with the functional boundaries of local communities. In this milieu, rural service centres were dense on the landscape and there was considerable uniformity in what was found in particular places. In late nineteenth-century rural Ontario, for instance, virtually every village possessed at least one general practitioner (Joseph and Bantock, 1984). However, in the industrial society, the wholesale acceptance of the telephone, good roads and the automobile 'revolutionised the life of country people and hastened a new rural civilisation' (Fiske, 1912: 66). These technologies weakened the local focus of rural life and simultaneously encouraged and permitted the consolidation of services into fewer settlements.

In both Canada (Fuller, 1994; 1997) and New Zealand (Joseph and Chalmers, 1998), the first 60 years of the twentieth century witnessed the gradual consolidation of services away from villages and small towns into larger centres characteristic of the industrial phase, and this was driven primarily by the impact of technology on three fronts (Joseph and Chalmers, 1998). First, changes in the nature and organisation of commercial agriculture and food processing resulted, cumulatively, in fewer workers (and their families) on farms and in food processing. Second, in both the public and private sectors providers sought greater economies of scale through consolidation of their activities into larger settlements. Third, automobile-based mobility allowed consumers to bypass local service providers in order to purchase cheaper or 'better' goods from distant suppliers. In some rural areas, de-population and loss of services seemed to be locked in a recursive embrace amounting to a 'downward spiral'. Fewer people meant poorer services and poorer services made it less and less attractive for people to 'stay on' (Bedford and Heenan, 1987). Work by Mermelstein and Sundet (1998) in the United States suggests that rural communities there followed a broadly similar trajectory of population and service decline. It is important to note here that the strength of this recursive embrace flowed from the fact that loss of services meant more than just having to travel somewhere else to obtain goods and services. Important though the economic and social costs of increased travel might have been, service depletion also meant the loss of jobs, and closures may sometimes have even threatened the very basis of local community identity (Kearns and Joseph, 1997).

In both countries, the settlement hierarchy characteristic of the industrial society began to give way in the 1970s to the more complex nesting of services and activity spaces characteristic of the open society. Mobility, especially through automobile ownership and use, was increasingly becoming a requisite of rural life. This greater expectation of mobility to take advantage of non-local service opportunities emerged in parallel with growth in the numbers of elderly people in rural communities, a cohort that had historically displayed high levels of transport dependency (Joseph and Fuller, 1991). Local governments and service agencies attempted to ease the impacts of technologically driven transitions on the elderly by subsidising local services or by providing assistance with transport (Joseph and Fuller, 1991). However, rural seniors remained conspicuously vulnerable, both to change itself and to failed interventions, especially of transport or mobility initiatives built on a narrow base of local voluntarism (Joseph and Martin-Matthews, 1993). In comparison to Canada, government support of rural servicing in New Zealand appears to have been more consistent and substantial through the 1960s and 1970s (Joseph and Chalmers, 1998), perhaps as a consequence of the more prominent position of agriculture in the rural economy (Joseph, 1999). While arguably lagging behind Canada in the transition from an industrial to an open settlement system in the early 1980s, New Zealand subsequently 'leapfrogged' Canada and other jurisdictions by embracing rapid and pervasive social and economic restructuring in the 1980s and 1990s (Kelsey, 1997). We now turn to this short cycle of change, with its inherent demand for rapid adjustment by all rural residents, regardless of age.

Short cycles of change

Britton *et al.* (1992) and Le Heron and Pawson (1996) have documented the geographical parameters of economic and social restructuring in New Zealand. In 1984 a series of changes was initiated that redefined the role of the state in the lives of New Zealanders through the overhaul of the social welfare and education system, the rationalisation of local government, the removal of state subsidies to agriculture and the sale of state assets and enterprises (Le Heron and Pawson, 1996). Social and economic restructuring hit rural New Zealand hard. First, the removal of subsidies and the opening up of agriculture to global competition forced rapid adjustments at the enterprise level (Cloke and Le Heron, 1994). The various adaptation strategies shared an important characteristic – they reduced demands for services, either indirectly through the displacement of farmers and their families (Collins and Kearns, 1998) or directly through reduced demand for labour and material inputs and for specialised services (Wilson, 1995).

Second, rural New Zealand was particularly hard hit by a retreat from the welfare state driven by the growing conviction that public services should be delivered in a 'business-like manner', a trend which a decade later began to reshape rural Ontario. Across the board, rural communities in New Zealand experienced the contraction of the public sector but not the corresponding expansion of private provision. Not surprisingly, private providers found it difficult to deliver services at a profit in situations where efficiencies in service provision had proved elusive to public providers (Joseph and Chalmers, 1998). Tirau, a small rural service centre of 750 people in the Waikato region of New Zealand, presents a stark example of service depletion. Because of its location on road and rail networks, the productivity of surrounding agriculture, and its status as an administrative centre for Matamata County, in the early 1980s Tirau possessed a mix of retail, professional and public services. However, in the late 1980s Tirau's fortunes changed dramatically, and for the worse. The year 1989 was a particularly bad one: Matamata County Council was dissolved and the new district council offices and attendant employment opportunities were located in a neighbouring community; the post office and bank were closed; and one of the local supermarkets, the local butchery, pharmacy and bakery went out of business (Joseph and Chalmers, 1998).

The net result of social and economic restructuring was a profound disturbance of the balance between rural population and services that had persisted, albeit sometimes precariously, through the middle decades of the twentieth century. In the 'new order' corresponding to the open phase of the arena society model's, it was assumed that residents of service-depleted communities like Tirau would travel to 'nearby' communities to obtain newly established or expanded (but often private) services (Joseph and Chalmers, 1995). It was also assumed that rural residents would, individually and collectively, accept their place in the new order (Joseph and Chalmers, 1999). In particular, it was assumed that frail or disabled elderly people would either manage to cope in their reconfigured communities or accept institutionalisation. Turning to the example of long-term care reform in Ontario, Canada, we see that equally profound changes in public (welfare sector) service provisioning were under way that were intended, at once, to help elderly people stay in their communities and to reduce government debt.

Prior to 1995, the long-term care system in Ontario had often been characterised as fragmented, uncoordinated and difficult to access; at best, it was a 'patchwork quilt of services' that allowed significant service gaps to exist in rural areas (Joseph and Martin-Matthews, 1993). The most often cited gaps have already been mentioned, including deficiencies in supportive housing, transport and mobility services, respite care, palliative care and mental health services (Hodge, 1993). In 1996, a reform-minded (Progressive Conservative) government sought to address the access and coordination problem by creating 43, one-stop, Community Care Access Centres (CCACs) across the province. However, instead of coordinating access to all institutional and community-based long-term care services, the CCACs excluded volunteer-based community services such as meals-on-wheels, respite care, telephone reassurance and friendly visiting from their funding envelope. This reinforced the reliance on voluntary-sector agencies and local governments for the provision of an important sub-set of community support services, and thereby perpetuated the systemic bias against rural communities exemplified by small, over-burdened volunteer networks and limited tax bases (Cloutier-Fisher and Joseph, 2000; Hodge, 1993; Troughton, 1999).

Cloutier-Fisher and Joseph (2000) illustrate the impact of long-term care 'reform' on the community care system in Ontario using the example of the rural town of Minto. Operating a system of 'managed competition', from the outset the CCACs had a swift and profound impact on the service environment. By 1998, a mere year or so into the new regime, both hospital and community-based professionals felt that they were losing their ability to act compassionately on behalf of their clients as regulations became tighter and new standards and guidelines were strictly imposed. For frontline workers such as homemakers, reduction in mileage reimbursements and system-wide policies requiring more visits to client homes for shorter periods of time – measures apparently indifferent to the greater dispersion of clients in rural jurisdiction – had the same effect of raising concerns about the place of 'compassion' and 'quality of care' for individuals in the new system.

In Ontario (Cloutier-Fisher and Joseph, 2000), as in New Zealand (Joseph and Chalmers, 1999), it proved difficult to attract private providers to 'distant' rural communities offering small numbers of clients and guaranteeing higher costs of delivery. Moreover, for-profit providers generally managed to underbid non-profit providers in easy-to-service urban areas but shied away from hard-to-service rural areas like Minto, thus depriving non-profit organisations of the ability to cross-subsidise within their cost structures. The promise of managed competition to deliver more efficient care, while improving client choices, was proving hollow for rural service recipients (Cloutier-Fisher and Joseph, 2000; Keating *et al.*, 2001). Overall, rural service recipients now faced a more restrictive range of service providers than before the establishment of CCACs.

It has already been noted that, while economic and social restructuring had hastened the emergence of the open society in rural New Zealand, a process that had been more gradual in Canada, key trends in servicing – fewer but larger services centres and greater demands on the mobility of service users – can be traced to earlier, pre-restructuring time periods (Joseph and Chalmers, 1998). This said, the introduction of market forces into the provision of welfare services, and the re-casting of service users from 'clients' to 'consumers', stands out as a more radical departure from historical continuity, overlaying as it does questions of financial access upon more traditional questions of geographical availability (Joseph and Chalmers, 1999; Cloutier-Fisher and Joseph, 2000), thereby creating new possibilities for the creation and perpetuation of vulnerability.

Growing old in vulnerable places: a view from New Zealand

A consideration of the situation of the rural elderly in the open society provides an opportunity to bring into focus some of the critical points of contact between long and short cycles of change in the rural sector. As noted earlier, it is hard not to see the elderly as being severely disadvantaged on the whole by the transition into an open rural society which provides considerable opportunities but simultaneously demands transport and personal mobility and, increasingly, expertise in a range of technologies. Older persons are caught in a 'squeeze' of sorts. The coincidence of local service depletion and increasing demands for mobility imposes a double penalty on the elderly; they have lost the (probably limited) services once available locally but may be unable to access with ease (possibly superior) services made available elsewhere.

Joseph and Chalmers (1995) report on interviews conducted in 1992 with elderly residents of Tirau, the service-depleted community in New Zealand referred to earlier. Respondents were almost unanimous in their condemnation of service losses. As one elderly resident stated, 'we want services back in the village', while another was of the opinion that 'by taking services away from Tirau it's no longer attractive for older people to live here'. The same elderly rural residents were all too aware of the impacts of not being able to drive. As one elderly person stated, 'I would be devastated if I had to give up driving'; another explained, 'if I had to give up driving, I'd have to rely on others if I wanted to go out of Tirau. That takes away your independence, which is such an important asset'. However, the impact of technology extends well beyond the realm of transport (and specifically the ability to drive) highlighted in the arena society model. The transformation of rural banking in New Zealand serves to illustrate this broadening of the technological challenge for elderly rural residents.

The de-regulation of banking and the privatisation of the state-owned Bank of New Zealand in the late 1980s prefigured a host of restructuring-led branch closures in New Zealand's rural communities (Britton *et al.*, 1992), including Tirau (Joseph and Chalmers, 1995). However, rationalisation continued through the 1990s in response to takeovers and mergers in the financial sector, but especially in response to innovations in electronic banking. Subsequent rates of branch closure eclipsed those associated with de-regulation in the late 1980s. In 1997 alone, 150 branches were closed nationally (KPMG, 1998: 41). The move to electronic transactions is underscored by the increasing use of automated teller machines (ATMs) and the growth of electronic fund transfer at point of sale (EFTPOS) outlets from 46,300 in 1996 to 59,992 in 1997 (KPMG, 1998: 26). Branch closures removed an important point of social contact in rural communities. Alluding to a colloquial term for ATMs in Australia, Argent and Rolley (2000) note that 'the hole in the wall has left a hole in the community'. For elderly people, who have on the whole found it difficult to adapt to new banking methods, service automation (in retailing and in interactions with government as well as in banking) based on information technology has created the potential for 'technological disability'.

In this discussion of the elderly in the open society, the partial convergence of what were once considered longer-term trends with short-term 'shocks' (captured under the rubric of economic and social restructuring) has been emphasised. In one sense, the flow of history, the very passing of time itself, transforms individual short-term shocks into what may at least in retrospect be seen as a cycle of change. In another sense, however, the nature of what were once considered long cycles of change may have been irrevocably transformed. This is especially true in the area of technology, where innovation and adoption cycles are now arguably more rapid than those once associated with economic and social restructuring. Given the pace at which business and government is embracing the Internet as a source of efficiency in service delivery, it is reasonable to speculate that in the post-open society which may emerge in the coming decade, mastery of information technology may for the rural elderly be as important as, or perhaps even more important than, the availability of automobile-based mobility.

Staying on in vulnerable places: a view from Ontario

Declining personal health and coping capacity alters the service needs of many frail and disabled older persons and obliges them to engage with the service system in different ways, and it is this need to renegotiate community support that renders this sub-set of elderly people particularly vulnerable to change in community-based long-term care systems. Cloutier-Fisher and Joseph (2000) use the 'stories' of eight elderly people in the town of Minto who combine poor health with difficult personal circumstances to bring to life the notion of vulnerability to restructuring. We draw on five of these stories here.

Meg, aged 73, experienced osteoarthritis, hypertension, diverticulitis and depression, along with what she described as 'poor nerves'. Meg indicated that she relied on her son to reduce her sense of social isolation, because visitors were so few nowadays. While her son acted as one of her main supports (and source of transport mobility) and although he lived nearby on a farm, he had two jobs, one of which (as a truck driver) often required him to be gone for several days at a time. Meg's perception of her community seemed to foster her sense of loneliness, loss and isolation as she remarked, 'so many stores are closed or are closing . . . you can't even get tea towels in town – you can't get anything you need'.

Joan was 73 years old and widowed. She had two sons, both living over 100 km away in different towns. Both boys were divorced and Joan said that consequently they didn't visit often, nor does she see her grandchildren regularly. She described her health as 'poor' and her financial situation as 'not very comfortable', noting that 'when my husband was alive, we could keep things going. Now, it is hard to keep up with everything'. Joan had gone through hip replacement surgery and suffered from osteoporosis and emphysema. She also had difficulty with her balance. While she did not expressly indicate that she had trouble with depression, she had multiple health problems, had given up many activities and had no immediate family to rely on, making her vulnerable to social isolation. Also, with her limited financial circumstances and the lack of available transport services in town she relied on friends or neighbours to take her to appointments or get her out of the house.

Meg and Joan represent considerable vulnerability, and at a level which challenges the notion that elderly people, their family and community can 'take up the slack' resulting from the partial withdrawal of the state from community-based long-term care. The fragility of arrangements is especially evident in the case of Ethel and Edward. Edward, a diminutive man with diabetes and circulation problems himself, greeted the prospect of reduced homemaker support for his 90-year-old wife, thus:

> Well, they said they were going to cut us down to one hour once a week. And, I said, well, I can't see how you can do that. Ethel needs a bath twice a week and I can't do it. And she [the case manager] asked, what are you gonna do? And I said, well, I'm going to have to see about getting help. But, I have to pay $14.00 an hour. They want me to do the vacuuming. Well, I said, I'll do what I can and what I can't do won't get done.

Edward's words capture some of the perceived implications of service withdrawal for ageing in place, where the outcome will be that people will struggle on without service because they cannot afford to do otherwise. This attitude has more in common with surrender and acceptance than it does with coping. Over time, an inability to cope with some of the major changes in community support systems will also increase the likelihood of institutionalisation for some. In this regard, two particular stories stand out among those recorded by Cloutier-Fisher and Joseph (2000), one concerns Joan, who went into a facility on a temporary basis because there was no proximate family member who could look after her when she broke her hip. Joan complained that during the short time she was institutionalised a number of her clothes were lost in the laundry. She ended up negotiating with her friend do her laundry for her. A second story concerns Annie, a 91-year-old woman who was wheelchair bound and afraid of falling during the times when she had to stand to put her supper on the stove. Previously, Annie's home-support workers put her supper on for her, but now their hours had been reduced and she had to fend for herself. She retold several horror stories about what she had heard about life in institutions and ended by saying, 'I'm not going anywhere, I'm staying in my home.' However, shortly after the interview Annie fell while making supper and was institutionalised in a town outside Minto.

Conclusion

The continued relevance of rural as an analytical category in the study of ageing seems assured, on at least two grounds. First, the existence and nature of advanced levels of population ageing in many rural communities, whereby the elderly constitute a high proportion of the total population but are relatively small in absolute number, distinguishes them from urban settings. In the latter, absolute numbers of elderly people are relatively large, but their proportionate presence is low. Thus it is not surprising that long-term care and other services targeted to the elderly continue to be focused on the easier-to-service urban settings in which the majority of elderly people reside (Rosenberg and Moore, 2001). Second, the rural community context for ageing has been transformed more radically than has its urban counterpart, and in the rural sector community attributes such as small population size, distance from urban centres, loss of local employment, and centralisation of government functions and decision-making outside the local community continue to promote vulnerability (Cloutier-Fisher and Joseph, 2000). The rural sector has been especially vulnerable to the impacts of technological change, as summarised in the arena society model, and while economic and social restructuring did not always target rural communities, the latter seem to have borne a disproportionate share of service losses and seen few of the gains.

What then of elderly people growing old in these vulnerable rural communities? We have already observed that they will have company and that they will be expected to help each other more than before (Joseph and Martin-Matthews, 1993; Bagilhole, 1996). It is also probable that they will continue to see their communities change, sometimes for the better but often for the worse.

For example, returning to the case of Tirau, Bedford *et al.* (1999) describe its commercial success in the late 1990s based on antique shops and other types of tourist-friendly retailing. However, they also observe that the new shops, restaurants and motel, while enhancing employment opportunities in the community, have not brought back with them the basic services (pharmacy, post office, bank etc.) lost a decade earlier. Indeed, some elderly residents had a rather jaundiced view of the 'new' Tirau: 'Tirau only caters to the tourists . . . the town changes haven't really affected the locals, only the passers-by' (Bedford *et al.*, 1999: 83). Indeed, drawing on the same data, Panelli *et al.* (2003) argue that the identity of the community has been remade in a manner that explicitly promotes the business community and implicitly marginalises the elderly, Maori and others.

The Tirau example suggests that it may indeed be impossible to 'turn back the clock'. The cumulative and mutually reinforcing effects of economic and social restructuring and of long cycles of change in rural population and service systems have created a strong momentum for change. This momentum is, in turn, being fed by new or reinvigorated forces of change, especially in technology, which now pervade most aspects of rural life. Yet most older people seem to want to continue to live short-distance lives in an increasingly open rural society that expects engagement with technology in more and more aspects of daily life. For example, even among computer-literate elderly rural people, the Internet appears to be used primarily as a supplement to other forms of social communication, and only rarely to obtain information about services (Fisker, 2003). Thus, while innovations such as telehealth hold promise for isolated rural communities, it is probable that most of their elderly residents will be dependent on others to facilitate their access to these space-shrinking technologies.

The local state, whether it is thought of as 'government' or as an extension of community, will continue to struggle with the competing needs of various age groups (Gitterman, 1991:Teather and Argent, 1998). Moreover, as noted in Canada (Cloutier-Fisher and Joseph, 2000) and New Zealand (Joseph and Chalmers, 1999), the de-regulation trumpeted in the drive to reform has invariably translated into a degree of re-regulation amounting to the virtual emasculation of the autonomy of local hospitals and other caring institutions. Much of the support for the least fortunate and most vulnerable individuals will therefore need to come predominantly from elderly people helping themselves or each other, or from their families. Cloutier-Fisher and Joseph (2000) cite the example of an elderly widow with breathing problems and osteoporosis who baked pies for a neighbour in exchange for yard work. In some cases, such as the example of Edward and his wife Ethel cited earlier, the choices vulnerable older people face may simply be to 'do without' needed services and supports.

We would like to think that the dual concept of vulnerability as it applies to small communities and to the experience of rural ageing, whereby elderly people are at risk because of their age and because of their location, is a transferable one. To some extent, we believe our dual focus on Canada and New

Zealand is already suggestive of transferability. We also see in the growing literature on ageing in east and south-east Asia further indications of ongoing transferability. For example, Joseph and Phillips (1999) provide examples of considerable variation in locally funded pensions and services for the elderly in rural China rooted in the differential economic success of local (often government-owned) enterprises.

In closing, we would like to emphasise again the importance of place in determining the experience of rural ageing. The situated experience of ageing flows in part from the constituent social, spatial and structural forces that shape, over the long and short term, the nature of rural places. For elderly people, and especially for those made vulnerable by illness and disability, life may become more and more focused on place; indeed, space eventually collapses into the confines of home (Joseph and Martin-Matthews, 1993). We have to be watchful for socially disadvantaged groups such as the elderly living out their lives in rural environments (Cheers, 2001: Merrett, 2001); careful that such home places and daily life environments do not close into what Dear and Wolch (1989) have elsewhere referred to as a prison of space and (lack of) resources.

11

AGEING AND THE URBAN ENVIRONMENT

David R. Phillips, Oi-Ling Siu, Anthony G.-O. Yeh and Kevin H. C. Cheng

Introduction: the local urban environment

The local environment is likely to be crucial to older persons, arguably even more so than for younger people, being the milieu in which they live and which therefore creates or hinders opportunities for ageing in place. In urban settings in particular, the physical townscape is likely to be of great significance to older persons and underpins their potential quality of life. It is also nevertheless important to remember that age per se is not the only issue, as an urban environment that is friendly for older people is also likely to be suitable for other age groups, too. Therefore careful environmental design will benefit all citizens. The Second World Assembly on Ageing held in Madrid in 2002 decided as one of its three priority directions the ensuring of enabling and supporting environments for older persons and their families (United Nations, 2002a). Within this priority direction, the Madrid International Plan of Action on Ageing identifies housing and the living environment as the first issue.

This chapter provides an overview of some of the main issues that have been identified with respect to barriers and difficulties mainly in the urban environment, although the general features apply to most wider environments. It is also important to remember that, as noted in chapter 4, the majority of older persons will not have handicaps and disabilities but that, permanently or temporarily, as age progresses, many may find some physical activities inside and outside the home more challenging. It is therefore almost inevitable if somewhat unfortunate that the debate is cast in terms of older persons as 'special populations', with disabilities or special needs. In many ways, older persons, especially in poorer urban environments, have been neglected in research terms (Slaughter, 1997), although there have been some studies and anecdotal evidence on, for example, the effects of supposed high crime rates on older people's well-being. However, comprehensive studies are rather fewer, perhaps because of the variety of people and environmental conditions met.

It is well to remember that to categorise any population by age or some other characteristic and expect all to be identical in their needs and behaviour is unwise. The individuality of people's needs in and reactions to their environment was well articulated by Golant in his 1984 book *A Place to Grow Old.* He also noted, however, that, in environments distinguished by predictability and certainty, human activities are less likely to yield spontaneous experiences. Many urban environments have quite extreme properties, both in their built forms (roads, high-rises, steps, gradients etc.) and social uses (noise, crime, crowding etc.). In this respect, the environment can continually or intermittently restrict and influence activity, as noted in many chapters of this book. Typically, an hierarchy of spaces have been used to describe urban environments. These include the wider townscape, the immediate local external environment outside people's home settings (these tend to be public or semi-private space) and the internal home environment, the domestic living areas of individuals and their families' private space. The debate is often cast in terms of access to opportunities.

Accessibility and barriers in urban environments

Recent studies of spatial behaviour have noted the importance of, say, civil rights and accessibility legislation in influencing the ways in which certain groups of people, sometimes called 'the other' or minorities or, more appropriately, 'special populations', are treated and how the environment is developed with them in mind. Imrie (1996), Swain *et al.* (1993) and Golledge and Stimson (1997) discuss how special populations may face problems and how some of these can be ameliorated and exacerbated. While older persons are definitely no longer minorities in many populations, several of their members have characteristics which make them special. In particular, Golledge and Stimson note that disabilities of various kinds have often traditionally been thought of in terms of problems of physical mobility. They emphasise that in reality disabilities can include many other problems, as well as the need for assistive tools such as crutches, walkers or wheelchairs. These can be vision impairment and blindness, speech impairment or language deficiency, hearing impairment or deafness, haptic (touch) impairment, cognitive impairment, reading impairment, phobias, mental challenges and debilitating diseases.

In epidemiological terms, as noted in chapter 4, older populations often have greater incidence and prevalence of many such areas of disability (Ebrahim and Kalache, 1996) and, without adequate social and physical design in the urban environments, these can become handicaps and cause considerable disadvantage. People with single and multiple impairments are often implicitly and explicitly discriminated against in terms of mobility, movement, access to services, employment, socio-cultural behaviour, education and information. This can be especially so in countries without equal-opportunity legislation and in many developing countries where fatalistic attitudes and superstition with regard to explanations of disability are widespread.

Table 11.1 shows some of the physical and human interaction barriers identified for disabled people by Golledge and Stimson (1997). They also note that

Table 11.1 Potential physical barriers in the urban environment

Visually impaired	*Age; phobic*	*Wheelchair users*	*People with other physical disabilities*
Construction/repair	Stairs	Construction/repair	Construction/repair
Weather	Door handles	Surface textures	Weather
Lack of railings	Lack of rails	Layout	Unprotected natural environments
Ramp availability	Door latches	Gradient	User access
Elevators	Elevator	Design features/ utilities	Stairs
Distance	Door closures	Unprotected natural features	Lack of rails
Doors – location, handles, latches, closures	Distance	User accessibility	Ramp availability/ curbs
Nonstandard fixtures	Ramp availability	Weather	Elevator
Traffic hazards	Distance	Stairs	Distance
Travel access	Elevators	Lack of rails	Doors
Surface textures	Lack of rails	Ramp availability/ curbs	Gradient
Overhead obstructions	Gradient	Elevators	
Lack of barriers	Unprotected natural hazards	Distance	
Lack of cues	Weather	Doors – width, handles, location, closures	
Gradient			

Source: after Golledge, R. G. and Stimson, R. J. (1997) *Spatial Behavior: A geographic perspective*, New York: Guilford Press.

there are very important issues which arise in terms of environmental design solutions which represent a 'therapeutic attack' on the environment, usually at the macro rather than the micro level, and tend to focus on physical rather than social remedies. A philosophical distinction is drawn between the American approach, which is typically to legislate for environmental design remedies in a self-help manner, as opposed to, say, the United Kingdom's social welfare culture, which looks for compensation, special treatment and pragmatism. American standards and specifications for developing barrier-free environments go back to the early 1960s and the broad-ranging anti-discrimination Americans with Disabilities Act of 1990. Comparable building and design legislation exists in Australia, Canada, the United Kingdom and many other developed countries. However, it is generally lacking or, where it exists, is ignored, in many developing countries.

The attention to detailed design in the immediate and wider environment for older persons and people with some types of disability is receiving increasing attention in many Western countries, not only because of the legislation requirements but also because of the increasing awareness of older persons as consumers – many of whom can afford to pay for the services and environments they wish for. It is increasingly recognised that relatively minor modifications to buildings at the design and initial build stage can be very cheap whereas to 'retro-fit' improvements is often much more expensive. Hence there is growing architectural and planning interest in creating 'inclusive' design features from the outset. This has been encouraged by the recognition that sometimes quite minor changes in design of, say, kitchens, bathrooms and walkways, can help avoid accidents and falls among vulnerable older persons and ageing populations (Lilley *et al.*, 1995). In Europe, the European Institute for Design and Disability (EIDD, 2004) stresses the philosophy of design-for-all, which aims to improve the life of everyone through design. Their website provides useful international links to organisations as well as legislation, many of which focus on design, disability and access to the built environment.

Distributions of older persons: urban and rural dimension

Worldwide, many older people are urban dwellers and, although some larger cities (especially certain Western capitals) have noted net losses of older residents in recent decades, many smaller and medium-sized towns have been gaining older populations (Warnes, 1994). On the whole, the distribution of the world's older populations is gradually mirroring that of populations as a whole and becoming more urbanised. This is in spite of the image of urban areas as being predominantly places that attract the young and migrants of economically active age groups. Urban growth and the ageing of populations in place have been largely responsible for the demographic ageing trends of most urban populations.

Today, very roughly speaking, developed-country older populations are about three-quarters urban while they are about one-third urban in developing countries. In 1990, about 73% of people aged 65 and over in developed countries lived in urban areas and this is projected to reach 80% by 2015. Corresponding figures for developing countries are expected to reach about 50% of older persons in urban areas by 2015 (Kinsella and Velkoff, 2001). Nevertheless, like all generalisations, this hides a range of differences locally, nationally and regionally, reflecting the particular patterns of urbanisation in different places. Older persons are generally more likely than non-elderly groups to live in rural areas and there are also gender differences. Elderly men are more likely to live in rural areas than elderly women, perhaps surprisingly, given greater female longevity, but reflecting past employment and migration patterns in certain regions of the world.

Residential distributions of older persons: the example of the United States since 1970

It has been reported in several United States studies that older persons' residential distribution in metropolitan areas is more similar to those aged 55 to 64 than those aged 20 to 9 (e.g. Golant, 1990). This suggests that while older people are not entirely segregated from the rest of the population, they are still the least likely group to share residence with younger groups of the population in suburban areas and, instead, they are closer to those who are similar to them in terms of age (Golant, 1990). There have been changes occurring over the past few decades, however. While nearly three times as many older people lived inside metropolitan areas in the United States in 1990, the percentage of non-metropolitan older groups was higher, at 15%, than that of the general population at 12.5% (Hobbs and Damon, 1996). There were some ethnic differences, with older Asians and Hispanics in the USA more likely to live in metropolitan areas than outside. The very old age group (aged 85 and over) was slightly more likely to live outside metropolitan areas than the younger-old group. Among older persons who lived in metropolitan areas, there was a roughly even split between those who chose to occupy central cities and those who lived in suburban areas. Despite these figures, it is reported that older persons are still more likely to occupy rural areas than the total population.

Future residential patterns of older persons may be dominated by ageing in place among the urban populations of the United States. However, some states are projected to gain older populations over the period 1990 to 2020. These include Florida, with its focus for inward migration, mainly to smaller and medium-sized towns, many of them coastal; Arizona, some Midwest states; and parts of New England. While ageing in place will underly a significant number of areas with rising older populations, there have also been influxes of older persons and the young retired for lifestyle reasons to places perceived as pleasant in the USA, as well as to many parts of other countries such as France and the United Kingdom. This emphasises the need for such receptor places to plan and forecast carefully likely services and infrastructure requirements of ageing populations who often have generally high expectations.

The urban environment: risks and benefits for older persons

Hazards in the internal and external housing environment

Hazards can potentially arise from features and aspects of the environment that pose a danger or chance of injury or harm to persons occupying or using the area. Some potential hazards arise from initial housing design and accessories inside the home, others can arise from poor maintenance or be exacerbated by the characteristics of the people occupying the spaces (Table 11.2). Features can include poor lighting, hard-to-reach switches, floor covering and rugs that are uneven or slip, unstable chairs or tables, chairs without armrests or with low backs, or unsafe electric appliances (Day *et al.*, 1994; Tinetti and Speechley, 1989; Lilley *et al.*, 1995). Bathrooms and kitchens have generally been identified as posing the most danger (indeed, bathrooms rather more so than kitchens, which have generally been thought to be the most dangerous room for older persons). In bathrooms, slippery baths or shower areas, toilets without grab

Table 11.2 Hazards in the home that might affect older persons

Types of hazard	*Description of hazard*
Inside home	Dim lighting, shadow or glare
	Light switches not clearly marked, cannot be seen in the dark
	Pathways not clear; small objects, cord, or tripping hazards present
	Carpet edges curling or tripping hazard; loose throw rugs, runners, mats, slip or trip hazard; area slippery, if non-carpeted
	Frequently used items stored where there is a need to bend over or reach up
	Step stool not sturdy
	Table or chair not sturdy, is damaged or moves easily
	Chair not sturdy, moves easily or needs repair
Kitchen	Light switches not clearly marked, cannot be seen in the dark
	Frequently used items stored where there is a need to bend over or reach up
	Step stool not sturdy; table or chairs not sturdy or move easily
Living room	Above general hazards with floor coverings; use of low chair difficult to get out of
Bathroom	Toilet seat too low or insecure
	Bathtub/showers slippery; lack of non-skid mat or abrasive strips and/or grab rails
Stairs	Dim lighting, shadows, or glare; lack of night lights; switches not at top and bottom
	Handrail not present, not sturdy, or does not extend full length of stairway
	Irregular height and spacing of steps and stairs; loose or damaged treads or carpeting
Outside	Sloping, slippery, obstructed or uneven pathways; slippery when wet; lack of rails
	General gardening hazards, especially involving tools or machinery

Source: based on discussion in Carter, S. E., Campbell, M., Sanson-Fisher, R. W. and Redman, S. (1997) Environmental hazards in the homes of older people, *Age and Ageing*, 26: 195–202.

rails and toilets located external to the house and other design features have been noted. Research in New South Wales found fairly high levels of risk in older persons' houses and, of 425 houses checked, fewer than 20% had no hazards. 42% had between one and five hazards, 26% six to ten hazards and an important 12% had between 11 and 36 hazards (Carter *et al.*, 1997). It is important to note that some design features of housing, while perhaps being suited for younger or fitter residents, may become less suitable for older persons. For example, round door handles may be used by people with adequate grip and strength whereas a person with arthritis and poor grip may prefer to use a lever handle. In addition, poor maintenance, not uncommon in some older persons' houses, may create hazards that did not exist before.

Design flaws and suggestions for improvement in the home environment

For many years the design of both living space and home appliances (ergonomic and safety features) has been studied by those interested in the welfare of older persons. In many countries, standards have been driven by general and specific (for older persons) building and architectural regulations and by safety codes encouraged in part by the consumer movement. However, only relatively recently have home appliances become widely available that are user-friendly to older persons. Unsurprisingly, problem appliances include ovens and stoves (design, placing and indication of controls), baths and showers (controls, access and slippery surfaces), counters (height), heaters (controls) and remote-control appliances such as for televisions (design of controls). Problems caused by intrinsic poor design are often exacerbated for older adults as users, who may suffer from physical impairments, poorer sight and tactile ability and perhaps the side effects of medication (Charness and Bosman, 1990; Charness and Holley, 2001). A significant minority will also suffer from some confusion and even dementia, especially among the very old population, meaning that housing and appliance design can make it impossible for them to live alone.

Some features can be overcome (for example, a step stool may be used in a kitchen to provide reach to standard kitchen cabinets for shorter people). However, for some older persons, climbing on steps poses a great danger from loss of balance which can lead to serious injury. Accident analyses suggest that grab railings might be added and steps might be widened. However, an alternative and perhaps a simpler initial solution is to lower the cabinets to suit the height of the users, although as a retro-fitment this would be very expensive. The problem could thus be addressed initially by installation of adjustable-height kitchen or bathroom cabinets, although this would be initially more costly and needs careful thought. Such systems are ideal for those in wheelchairs or unable to reach above head height. However, cost consideration of retro-fits are considerable, even with government subsidies, which can limit improvements to the well off or to those who have financial assistance from the government or charities. Clearly, in many developed countries, most older persons do not have the resources to make such home improvements and, in developing-country cities, the people able to afford such luxuries are even fewer. This is a major concern for most seniors who are limited by their financial situation.

Home design technology, including the application of information technology, is a major area of growing interest for designers in Europe, North America, Australasia and areas of Asia. These include numerous appliance safety refinements such as auto controls, cut-offs and temperature regulators as well as readable and usable instrument panels. Many local and international organisations, such as Age Concern and others, report on unsafe household items such as electrical products that older people might wish to avoid, or on appropriate items they might wish to acquire (Age Concern, 2003). As access to, understanding of and use of information technology is so crucial to all people's access to information, services and quality of life, SeniorWatch (2003), a European Union funded project, surveyed varied aspects of older persons' computer requirements and usage, noting that older persons represent some 20% of the population in Europe and are already often significant purchasers and users of IT. In the more specific assistive technology sector, information society technology (IST) products and services specifically aimed at meeting the requirements of older and disabled users are seen as particularly crucial for the future and form an important market. This is reflected in the EU's eEurope initiative, which identifies the huge potential to enrich everyone's life by bringing communities closer together or sharing knowledge.

At a more basic level, the importance of removing or at least reducing environmental risk factors for injury to older persons is imperative in ageing populations. As discussed below, for all types of accident covered in a review by Lilley *et al.* (1995), including falls, road traffic accidents and burns, older persons had a worse outcome than younger victims. The true prevalence, seriousness and long-term care costs associated with these accidents, as well as the loss of quality of life for the injured people and their families, renders this an area of great importance for further research.

Road traffic and transport hazards in the external environment

The wider setting in which the older person's home is sited forms the external environment (Marans *et al.*, 1984). Environments are of course very varied and include, as well as housing, green areas such as parks, gardens and common areas for recreation and rest, and passages, flyovers, subways, traffic density, air quality and surrounding noise from neighbours, traffic or construction work (Loo, 2000). Foremost among cause for concern is road traffic, in its hazards to both the older pedestrian and the older driver.

In many ways, transport is one of the key environmental areas to be evaluated and planned carefully, as it has the capability to affect social integration and provide access to life-sustaining functions. It therefore very much influences the ability of older persons to remain active and independent and to age successfully. The ability to travel or drive frequently becomes a principal factor determining whether an older person can remain living alone or will require long-term care and assistance. Some local environmental schemes can be helpful, including traffic calming measures, pedestrianisation and other traffic–pedestrian segregating schemes. These measures will, no doubt, become a crucial component in improving satisfaction and independence in later life.

Road traffic accidents are important to all age groups and many can be avoided with appropriate environmental design and traffic control measures (Evans, 1999). However, older persons injured in road accidents are three times more likely than younger people to die as a result of their injuries. Accident records from the United Kingdom show that more than a fifth of car drivers and almost half of the pedestrians killed were aged 60 and over. While there are relatively fewer older drivers and they have relatively few accidents, older drivers tend to experience more fatal accidents. Older drivers and car occupants are generally more seriously injured for any given crash exposure than younger adults and they are hospitalised for a longer time for a given injury and have more disabling injuries. With respect to pedestrian accidents, in spite of a general reduction in the number of casualties and deaths in most developed countries over recent years, the number of such accidents involving older people remains higher than in other age groups. Therefore road schemes, signage and vehicle design that assist older persons will be very important and will of course benefit all road users. For all ages, safe location of bus stops, easy-access public transport vehicles, accessible subways and walkways fitted with escalators and elevators are essential. These all need to be provided bearing in mind the common environmental hazards identified in this chapter.

There have been relatively few detailed studies highlighting effects on older persons themselves of transport systems and transport improvements. One project reported surveys taken before and after pedestrianisation and improvement schemes in Kingston upon Thames, London, which were found to be generally well accepted (Smith, 1994). Nevertheless, significant problems remained for some disabled and older persons although numbers experiencing difficulties with narrow or obstructed pavement walkways had decreased. Special road crossing facilities such as pelican crossings (which emit sounds when the traffic lights are safe for pedestrians) can be useful but can also add to confusion and panic among slower-moving pedestrians. Likewise, some communities in countries such as Japan have installed sidewalk guidance indicators in the pavements but many older persons have found these to be confusing. However, as drivers, older persons may have somewhat slower responses, and perhaps poorer eyesight. Recent research suggests, however, that older people with driving problems may successfully use compensating strategies (such as speed adaptation, anticipatory behaviour, sticking to familiar routes and avoiding rush hours) to avoid accidents (De Raedt and Ponjaert-Kristoffersen, 2000).

Falls and injuries among older persons

Falls are a leading cause of injury and death in people aged 65 years and over, accounting in the United States, for example, for one-third to two-thirds of all accidental deaths (Lord and Sinnett, 1986) and similar figures are reported for Australia (Day *et al.*, 1994). Studies have found that one in three of those aged 65 years and older and living in the community falls at least once each year (Tinetti and Speechley, 1989). Falls and injuries generally can be much more serious among older than younger persons for physiological reasons. In addition to injury from falling, falls can result in prolonged recovery, loss of mobility and even death because of circulatory problems. Falls can have considerable psycho-social impacts on older persons and can result in restriction of activity, a fear of falling and a reduced quality of life (Vellas *et al.*, 1989; Tinetti *et al.*, 1990; Lilley *et al.*, 1995). Even falls that do not result in physical injuries can result in a post-fall syndrome, with many older persons experiencing loss of confidence, hesitancy and tentativeness, with resultant loss of mobility and independence. Indeed, after falling, as many as 48% of older persons report a fear of falling and 25% report curtailing activities. Tinetti *et al.* (1995) also found that 15% of non-fallers reported avoiding activities from a fear of falling. In a prospective study of fallers and non-fallers, Vellas *et al.* (1993) investigated the consequence of falls among small samples of both community-dwelling and institutionalised older persons. They concluded that falls resulted in restrictions of activity for six months after the fall and suggested that falls appear to be a factor which aggravates and accelerates the effects of ageing.

Causes of falling and fall risk in the urban environment

The epidemiological importance of falls as a cause of injury and death in older populations being so great, it is important in the planning and design stage of housing and urban schemes to try to avoid the main known features that increase the incidence of falls. To 'design out' such dangers, it is essential to know why falls occur and under what circumstances they take place. The risk of falling clearly occurs when an individual engages in an activity that results in a loss of balance, a displacement of the body beyond its base of support. However, the environment can also be a key variable.

Falls in older persons are often precipitated by a number of factors. Lilley *et al.* (1995) note that these encompass intrinsic factors (for example, age-related physiological changes, disease states, medications) and extrinsic factors such as hazardous environmental conditions, faulty devices and footwear, especially slippers (see also Lach *et al.*, 1991; Campbell *et al.*, 1981; Tideiksaar, 1989). Loss of balance may occur while carrying out everyday activities such as walking; sitting and standing; getting into or out of wheelchair, bath, bed and toilet; and reaching up or bending to retrieve or place objects. Estimates suggest that between one-third and a half of all falls in older people in the community are due to environmental factors. These include uneven or slippery floor surfaces (including the presence of rugs and mats), tripping over obstacles (including pets), inadequate lighting, poorly designed or maintained stairs without handrails, inappropriate materials used in flooring and stairs, and inappropriate furniture design. Such features alone or in combination often increase the risk of falling, tripping or slipping for older persons. Stairs may contribute to a number of serious falls but the majority of falls reported by people over 65 occur on the same level, suggesting that other factors are important.

The location of falls

Within the house, the major hazardous areas for falls among older persons include bathrooms, kitchens and dining areas (Carter *et al.*, 1997; Lord *et al.*,

2001). It has been found that over 80% of homes have one or more hazards and multiple hazards were frequently found in rooms and areas where older people perform complex daily routines (such as showering or washing in the bathroom, cooking in the kitchen) or which require complicated motor actions (such as climbing stairs or getting on or off the toilet).

Around half of all falls experienced by healthy community-dwelling older persons occur within the person's own home. Campbell *et al.* (1990) found that 16% of falls occurred in the person's own garden, 21% in a bedroom, 19% in the kitchen and 27% in the lounge or dining room. Other studies have found 6% of falls occurred while using the shower or bath, 3% were off a chair or ladder, 6% on stairs and 26% while walking on a level surface (Lord *et al.*, 2001). This shows that many falls occur while people are carrying out daily routine tasks. It also clearly highlights the potential influence of environmental and design factors in gardens, in public places and in other people's homes, where familiarity with the setting may be reduced, in addition to those risk factors within the person's own home.

While numerous environmental factors have been implicated as possible causes of falls among older persons, there is relatively little hard evidence that environmental factors are the primary risk factors for falling (Lord *et al.*, 2001). For example, it has not been conclusively demonstrated that older persons who live in more hazardous homes have more falls than those who live in safer homes. The lack of associations may reflect, at least in part, the difficulty of studying transient or intermittent risk factors and of accounting for people's familiarity with their own home environments, or transient ill-health episodes. However, since, as mentioned, one-third to half of all falls do appear to involve or be exacerbated by environmental factors, it seems that the interaction between older persons' physical abilities and the environment is a crucial factor in determining whether a fall will occur and perhaps how serious it can be.

Older persons' basic needs and the environment

In general, older persons have the same basic needs and aspirations as everybody else, for example, they aspire to be healthy, well fed and housed, and financially secure. They want a home that is adequate to their needs, a place that is warm, comfortable, affordable and secure (Robson *et al.*, 1997). They also desire privacy, independence, friendship, love and dignity. However, many older persons have some additional needs and features that may slightly or considerably alter their priorities. As their dependency level may increase, so they may come to rely more and more on other people – relatives, friends, neighbours and professional carers – to assist them with daily tasks and to meet their basic needs. Therefore access to such sources of assistance becomes very important, as emphasised in the Madrid International Plan of Action on Ageing (United Nations, 2002a).

The notion of autonomy and security in older persons' living environments

The goal of designing a supportive environment should arguably be to maximise a person's competence or ability to function within their environment, whether in a building, a neighbourhood or the wider city. Environmental competence is generally held to be a function of two components: someone's personal capabilities and the characteristics of the environment itself. Therefore it is highly unlikely that a supportive environment can be designed without first understanding the needs and capabilities of the likely inhabitants and also how these characteristics will change over time (for example, as a population ages).

The concept of a basic human need for competence was first elaborated in depth by White (1959), who discussed the need to feel competent in transactions with one's environment as a central motivational force underlying much of human behaviour. Empirical research and theory have emphasised that older persons need to maintain perceived and exerted independence especially as they move from private residences into, say, special housing environments (Parmelee and Lawton, 1990). Autonomy and a sense of control are important in that minor increments in control of personal activities and social contacts can give substantial positive impacts while the loss of such control can have negative impacts. Indeed, perceived lack or external locus of control has been associated in a range of studies over the years with poorer adjustment, activity and physical health among older persons living in a range of settings, from independently in the community to nursing homes and congregate housing. This factor can certainly affect the ease of transition between private and institutional living.

Like social theorists, housing designers and architects increasingly recognise the importance of autonomy, but their practical work places equal if not greater emphasis on security as a primary need. The term 'autonomy', as used in the gerontological literature, spans a variety of conditions presumed to contribute to (or, by their absence, detract from) freedom from risk, danger, concern or doubt – that is, to enhance both physical safety and peace of mind. Both empirical studies and practical design guides emphasise the importance of visual orientation, reduced risk of accidents, environmental familiarity and neighbourhood integration as contributors to older persons' physical safety and psychological well-being in community and congregate settings (Howell, 1980; Hunt and Roll, 1987). The following brief discussion outlines some guidelines for the environment that may enhance living conditions for older persons.

Designs and housing arrangements for older persons

Many authors note that positive and negative physical aspects of the housing environment can influence an older persons' activity level, social contacts, well-being and general lifestyle. Moreover, the physical attractiveness of housing has been related to tenant satisfaction (Carp, 1978), and better physical facilities have been associated with enhanced resident functioning (Lawton *et al.*, 1978).

Services and facilities provided and accessible, physically and financially, are clearly of great importance to residents' well-being. In studies probing the services desired by public housing residents, medical services are generally accorded top priority (Carp, 1976; Lawton, 1975). Sherwood *et al.* (1981) showed that even though off-hours nursing services were not extensively used in a public housing facility, a proposal to eliminate them was adamantly resisted. Residents felt reassured that assistance was available if needed. Other services important to residents for maintaining their independence include transport to medical service providers; assistance with chores and housekeeping, meal preparation, cleaning, shopping; personal assistance (such as in bathing, clipping toenails or doing laundry); day or night care; care for some prostheses; and daily monitoring to ensure that the residents remain well. Social, recreational, educational and cultural provisions are also widely held to be desirable.

Planning issues

For developers, legislators, politicians, town planners and other stakeholders (including the residents) in the official aspects of environmental planning and accommodation design, numerous factors need to be considered when proposing changes to the environment where older persons reside.

Older persons' habits and constraints

As noted earlier, a large percentage of older persons live in urban areas and most older urban dwellers today have 'aged in place' (*in situ*), having lived in their current dwellings for a long time, often for several decades or more. The newcomer who has migrated into an area on retirement is unlikely to be typical of most older urban residents in the future, except in a few favoured retirement location settings (Phillips, 1999). Therefore the suitability of the residence and the environment may well change as people's needs and abilities evolve. Many older persons can face deterioration of their property, which may indeed be a family home now too large for them if their children have left, and they may find it difficult to maintain it due to rising costs and their own relatively fixed incomes (Golant, 1984; Phillips and Yeh, 1999). If they want to remain in their neighbourhood there may also be other problems. Older persons generally have established patterns of shopping and daily activities in the neighbourhood (Myers and Huddy, 1985; Phillips, 1999). However, many older persons do not drive or no longer drive (Phillips, 1999; Lawton, 1972) but rely on walking, public transport, or family, friends and volunteers to transport them. They often have local activity spaces, partly caused by choice and familiarity, but also sometimes constrained by their inability to journey far from their residence. They can thus be denied or restricted in their access to certain goods and services that are available in the wider urban community but not locally. Their choice of services can be constrained and they may be forced to satisfy their needs in a relatively circumscribed neighbourhood.

This emphasises that people's needs and perceptions of their environment are not static but will change especially as they age (Walsh, Krauss, and Regnier, 1981; Golledge and Stimson, 1997; Phillips, 1999). If urban renewal or regeneration projects cause local shops to close or move, as sometimes happens, older people may find their access and service choice reduced yet further or even eliminated altogether. For those who are physically disabled or of very restricted mobility and income, the problem is even more severe.

Building designs for handicapped and older persons

Physical and architectural features

It has been widely acknowledged that positive physical activities of the housing environment can influence an older person's activity level, social contacts, well-being and general lifestyle. This has recently re-emerged as a major area of concern in gerontology, after some years of lesser attention (Maddox, 2001). The physical attractiveness of housing has been related to tenant satisfaction (Carp, 1976), and better physical facilities have been associated with enhanced resident functioning (Lawton, 1975; Linn *et al.*, 1977).

The environment can be analysed by looking at the presence of physical features that add convenience and special comfort and foster social and recreational activities (physical amenities, social–recreational aids). We can also look at features and facilities that aid residents in activities of daily living and in negotiating the environment (such as prosthetic aids, orientation aids and safety features). In addition, the extent to which the physical settings offer residents potential flexibility in their activities (such as architectural choice, space availability) is also crucial, especially as older populations in the future are much more likely to be community-dwelling and diverse in their needs and abilities. Last but not least are factors related to the community itself, especially the degree of physical integration between facilities and the surrounding community, a form of community accessibility.

Stress and coping

Stress has several definitions. Some define it as a combination of physical, mental and emotional feelings that result from pressure, worry and anxiety. Others refer to it as feeling which results from excessive pressures and environmental stimulation. Stress among older persons is a human reaction, but not an intrinsic feature of the environment itself although there is clearly potential for the environment to increase or decrease the conditions for stress. Stressors (items that cause stress) can thus stem from various aspects of the social and physical environment. The issue of coping with stress may become increasingly relevant as chronological age advances since, among many factors, disabilities, declining health and losses such as of loved ones, financial support or roles are common among older-aged groups (Cheng *et al.*, 2002).

There are different ways to assess the effects of environmental stressors and some will be acute while others are chronic. Life stressors can include both episodic stressful events and chronic, ongoing strains such as failing health and hospitalisation, cognitive and functional decline, interpersonal losses and financial strain (Burnette and Mui, 1994) and stressors can vary in their predictability and controllability (Veitch and Arkkelin, 1995). A typology of stressors allows distinctions to be drawn along several major dimensions, including how long the stressor persists, the magnitude of response required by the stressor and the number of people affected (Evans and Cohen, 1987).

The category that includes some of the most chronic environmental stressors has been labelled 'daily hassles'. These are present during most of our daily lives and include numerous factors such as neighbourhood problems, crowding, noise, air pollution, congestion and many others. These assorted daily hassles, when they accumulate, can affect behavioural as well as psychological and physiological responses (Siu *et al.*, 2003). It may be that this will have a greater impact on older people than on those who are younger, if they are restricted to the local environment, as their exposures will be more prolonged. The assumption is (perhaps erroneously) that the ageing processes render some older persons less able to cope with the demands of environmental stressors.

Stress moderators

Individuals can adopt adaptive strategies, plans that are used in either stressful or non-stressful situations to obtain a desired outcome in the face of a given challenge or need (McCraig and Frank, 1991) (see also chapters 9 and 14). Such strategies can clearly help people to adapt to, or survive in, less than ideal environmental settings. For instance, if an older person is unable to go on shopping trips as a result of physical immobility, he or she may instead shop at home through the television, phone or Internet; solicit requests for salespersons to make visits to the home; ask a friend to shop for him or her; or use a combination of the above approaches. Hence the effective use of such adaptive strategies has the potential to mitigate the effects of chronic illness and thereby increase older adults' capacity to maintain stable and satisfying daily activities. In this regard, behavioural and lifestyle factors that predict healthy ageing, such as exercising regularly, abstaining from excess alcohol and avoiding tobacco, practising good nutritional habits and taking advantage of social support systems (Clair, 1990; WHO, 2002), can be construed as adaptive strategies. They are potentially important because of their possible role in bolstering feelings of self-efficacy and personal control, factors that are related to positive life outcomes in older adults (Holahan and Holahan, 1987; Reich and Zautra, 1984; Rodin, 1986). Key among the moderators of stress among older persons are a sense of personal control and also religious feelings and beliefs (Krause, 1998).

Recreational activities among older persons in urban environments

Stress has been related to a number of poor health outcomes, including depression and some physical illnesses and the undermining of normal immune system functioning. In response to stress, older persons may choose to participate in an abundance of leisure activities, including attending classes and clubs, walking, making gifts for others, spending time with friends, attending church, and praying. Other leisure activities also included listening to music, attending concerts, caring for a pet, reading, going to parties, performing magic, cutting flowers and having sex. In general, activities were found to be useful to enhancing quality of life and psychological well-being (Clark *et al.*, 1996), although of course many of the social theories of ageing are ambivalent on the actual nature of the relationships between activity, disengagement and role adoption. On a daily basis, nevertheless, there are many activities of daily living (ADLs; Baltes *et al.*, 1990), the basic tasks necessary for daily living, including eating, grooming, dressing, resting or cleaning. To continue these, many older persons use a variety of strategies to adapt to environmental demands (Clark *et al.*, 1996), including using technological aids or adaptive equipment, practicing good health habits and structuring one's environment to minimise the burden of performing necessary activities. Informal help from friends, relatives and neighbours can assist with the performance of required daily tasks as can formal services help in daily tasks (such as meals-on-wheels, home helpers to assist with cleaning tasks, home delivery of shopping).

The coping strategies adopted by older persons have been distinguished by the functions they serve and three concerns are commonly found to underpin them (Clark *et al.*, 1996).

Safety concerns may involve threats to physical and social well-being, violence (real or perceived), physical disasters and personal mishaps. Strategies include maintaining oneself in a safe physical location, using the protective advantage of other people and taking general precautions when out. *Living a satisfying and happy life* involves strategies such as keeping active by engaging in worthwhile activities, using formal professional services to remain happy or well, maintaining a positive state of mind and treating other persons properly. Finally, *relationships with others* can help in the environment and can be maintained and improved in a variety of ways such as by making use of formal gatherings to meet and interact with others, fostering positive relationships with others, reaching out to family and friends or finding substitutes. Last, but not least, avoiding others (disengagement) can mitigate some negative aspects of social life and is a fairly common way to cope with stress among older persons.

Conclusions: inter-sectoral and inter-disciplinary approaches to environmental planning for older persons

It is clear that, worldwide, with demographic ageing and ageing in place, older persons are going to form an increasing percentage of urban populations. Therefore enhancing living conditions, making lives easier and more enjoyable, is likely to become of increasing importance, especially as an ever larger proportion of older persons are going to live in the community (Mende, 2000). The growing interest in understanding the process of ageing has increasingly been incorporated in the multi-disciplinary field of gerontology. Many disciplines are concerned with many aspects of ageing, from studying and describing the cellular processes involved, to seeking ways to improve the quality of life of older persons. Among many others, the Gerontological Society of America increasingly emphasises inter-disciplinarity, which certainly should apply strongly to the planning of environments and design of accommodation suitable for all ages. As stated at the start of this chapter, an environment that is conducive for good living for older persons is also likely to be good for all ages, so this is yet more important if inclusive communities for all ages are to be developed.

There is clearly an imperative for inter-sectoral and inter-disciplinary approaches to comprehensive environmental planning for older persons (Phillips and Yeh, 1999). The ageing process occurs in a social context, which helps to determine the meaning of ageing as both an individual and a societal experience. Gerontologists are well placed to act as brokers in the inter-sectoral activity, particularly those with a training in or sensitivity to the importance of place. Gerontologists operate not only in academic discipline but also in a field of practice that involves many aspects of public policy and human service (Kart and Kinney, 2001). The disciplines and interests likely to be included in successful urban planning are manifold: planners; architects; housing managers; engineers; social planners; transport planners; environmental design specialists; the medical and paramedical sector, including social epidemiologists and social service representatives; and a range of academic disciplines, including health geographers and psychologists, and especially environmental psychologists, economists and policy specialists. Key is the inclusion of the user communities, older persons and their families themselves, although sadly the wholehearted inclusion of users in environmental planning is rare and their interests are sometimes represented only superficially by elected representatives and planning officers. Truly inter-sectoral planning is still very rare in the Western world where departmentalism and fragmentation are more the norm; in most developing countries, this is even more the case. However, in this chapter, we have attempted to introduce some of the major concepts and issues involved. The future lies in the implementation of many of the combined strategies identified.

12

IMAGINED LANDSCAPES OF AGE AND IDENTITY[1]

Andrew Blaikie

Why do we associate particular places with old age? It is, of course, possible to devise an inventory of the actual social settings in which older people find themselves and to explore the contouring, architecture, design and social structure of such real places – domestic interiors, city streets, holiday resorts, care facilities, long-stay institutions, bingo halls and so on – and to conclude that these have clear implications for stereotyping, both positive and negative. With Foucault to hand it is then relatively straightforward to theorise the spatialisation of age through discourses and disciplines (Katz, 1996). Some places are sites of struggle and resistance, others of accommodation and incorporation; some effect sharp boundary maintenance between second-, third- and fourth-age domains, others allow easy slippage across generations. Each lends meaning to the ageing experience. However, my concern is less with the places where people age – actual or epistemological – than with the imaginative salience of such places, with what they stand for in the mind's eye and the existential utility of those imaginings.

Introduction: the metaphorical landscape

Ageing occupies a strongly symbolic landscape in that there exists a powerful metaphorical association between chronology, place and space. Life is pictured as the journey of life; the life course is seen as a river, taking many bends and turns; life stages are pictured as a flight of steps ascending towards the zenith of early adulthood, levelling out, then descending after retirement; our days become a downhill slalom ride where each set of gates moves about bewilderingly as we approach them. We travel collectively in cohorts, like flocks of birds in flight, or shoals of fish along oceanic currents. The individual life course is portrayed as a lifeline that intersects with many figurations, and harmonises or jars against the developmental cycle of the family, career paths and so on. Infused with such imagery, which has a long historical pedigree, social thinkers have developed models to explain our travels – Hareven's temporalities, van Gennep's *rites de passage*, as the hoops through which we must leap to progress from one stage to the next; Glaser and Strauss's status passages; or, as populist synthesis of a range of psycho-social theories, Gail Sheehy's *Passages* pure and simple (Hareven, 1977; van Gennep, 1960; Glaser and Strauss, 1971; Sheehy, 1976). The world is not so much a stage as an orienteering course, replete with currents, passes and pinnacles to be navigated. Many will reach the crossroads, Dante's place in a dark wood, not knowing which way to turn (Dante, 1308). Yet we are driven on because the trajectory is uni-directional. We cannot go backwards. We are born dying and, for all we might indulge in memories, in longing or nostalgia, there is no ultimate means of denying our mortal destiny.

But if the ageing process is a journey, it also cuts a path through a terrain that is itself varied: standing on the plateau ridge of mid-life, both 'the blue remembered hills' of childhood and Lear's blasted heath of decrepitude seem far away. The past is a distant land, so too is the future: 'old age is not interesting until one gets there, a foreign country with an unknown language' (Ford and Sinclair, 1987: 51). This transposed sense of otherness sets apart old age as a far and separate landscape, and while gerontologists have been at pains to build bridges by forging friendly ties and adopting an anti-ageist common currency, they have been equally eager to maintain disciplinary borders by keeping their subjects self-contained. The 'scientific study of old age . . . and of old people's special problems' is not really the investigation of ageing as a process that permeates the life course (Sykes, 1976: 445).

In a manner reminiscent of Victorian exploration, social researchers have characterised the sociology of later life as uncharted territory awaiting their discovery – a dark continent like Victorian Africa. Few have been so bold or visionary as Peter Laslett, who suggested we threw away the old charts and unfurled 'a fresh map of life' in which the demography of whole nations defined a need for a different paradigm (Laslett, 1989). His life-course perspective implies a grasp of contingency – a realisation that what happens to one cohort of people at one point in time, and in their own lives, will have far-reaching effects both across contemporaneous generations and for that generation later in time. The dimensions are both horizontal (cross-sectional across social groups) and vertical (longitudinal, for that group over time). However, the designs of our metaphorical picture lack this three-dimensional contextuality. Instead, they are devised as pocket guides for individuals caught up in their own time. As one writer puts it:

> Our maps have improved in a kind of spurious precision as they have deteriorated in the amount of worthwhile information they convey . . . If the Ordnance Survey were worth the paper it prints on it would be producing a special series of maps for the over-forties, the over-fifties etc. (When you reach a hundred you end up merely with an enlarged plan of your house, with the location of the lavatory clearly marked and a small arrow pointing towards the crematorium.)
>
> (Maclean, 1986: 106)

While locally helpful, such plans firmly position us within age-appropriate zones. Older people, like everyone else, have their place. It requires a literary imagination to escape such consignment – and escape will be a key theme – hence, for example, Thea Astley's recipe for elder-anarchy in her novel *Coda*:

> It's time to go feral. Tribes of feral grandmothers holed up in the hills, just imagine it, refusing to take on those time-honoured mindings and mopping up after the little ones while the big ones jaunt into the distance.
>
> (Astley, 1995: 153)

The image is striking, not just because it positions old ladies as combat troops hiding in the hills, rather than as domestic drudges. In the scheme of things, it is not right because, as we know, the twilight scene should focus on a peaceful, rustic cottage with roses round the door. The popular media, and especially advertising, draw upon a conventional stock of caricatured images where the ageing body (the autumn of one's years) is chronologically complemented by signifiers of diurnal and seasonal change in the natural environment (sunsets and falling leaves). Accordingly, popular culture invites us to picture some scenes as peculiarly age-appropriate, others as singularly deviant.

Meanwhile, the symmetry of the journey of life (up the steps, along the mezzanine of mid-life, and down again) is complemented by continual reference to life as a cycle and, as a globe turns on its axis, so the life-cycle view sees life as returning ultimately to its source. Not only do we come from the unknown, thence to return, but deep old age comes to mirror early childhood, hence infantilisation – the reversion to 'second childhood' – being regarded as a logical end point (Hockey and James, 1993). Nevertheless, because for each of us life is also a journey, we have the opportunity to strike off down other routes, to take other life courses. Thus the symbolism of ageing may be largely that of social constructs that constrain us – that would certainly suit a Durkheimian interpretation – but our culture of individualism ensures that we regard ourselves as self-determining creatures.

Real and imagined worlds

Can we therefore cast the age-old problem of structure versus agency in terms that reflect the imaginative spatialisation of the life course? The realities of later life sometimes mask, as others project, a mental landscape of imagined alternatives. That is to say, some older people – especially younger, active third-agers – are able to develop lifestyles in retirement that realise their visions of an ideal world for themselves – what I shall call projection – whereas rather more people, and particularly those whose life chances are limited by reduced incomes or failing health, may envisage a better world but are less able to create it (Blaikie, 1999). I have termed this involution. What is interesting in each case is what people imagine and how this is done.

Two developments in the sociology of knowledge are helpful here. First, Berger and Luckmann compare the many imaginary realities we might inhabit daily – sleep, daydreams, drugs, sport, the cinema – with paramount reality, the absolute existence of the routine here-and-now (Berger and Luckmann, 1991). The fundamental difference lies in the fact that wherever we might take ourselves in our mind's eye, we cannot ignore the obduracy of immediate reality – it will not go away. Despite this, Cohen and Taylor argue that we need constantly to create alternatives – what they call 'escape attempts' – if we are to remain sane (Cohen and Taylor, 1992). They compare the routines of everyday life with the managed round of the prison. Like prisoners we resist workplace institutionalisation and boredom by creating other routines: reading, hobbies, games, fantasies, weekends, holidays. All are temporary reality slips, since ultimately we must return, hence 'attempts' rather than true escapes. Yet all sustain us in getting through our lives. Let us think then of ageing itself as the paramount reality: it is obdurate because it is inevitable, yet we can, and do, construct many escape attempts which may be fulfilling, even liberating, but are by definition fragile and temporary alternatives. Clearly, a spectrum exists: for the severely disabled nursing home resident, trapped in an institution, the alternative realities might be limited to television or reminiscence; but, for the sprightly, affluent retiree, the world cruise may beckon. Such alternatives are not randomly generated. In the immediate here-and-now, the mental and physical health of the individual and the material circumstances in which s/he is placed are critical. But there is also the cultural history that attends that individual's particular biography. If our identities derived solely from where we were at any one time, we would become extremely confused, not to say mentally exhausted with the effort of placing ourselves. Thus we construct for ourselves workable biographies, reflections upon our lives through which we interpret our ageing as it happens. Each of us needs to connect with our own past. At the same time we can only assemble our agency from the resources locally and presently available to us. What visions stem from this juxtaposition?

Projection – looking forward

Let us begin with projection. Assuming health and wealth, what can older people achieve when they access their cultural past as a foundation for building the present? A few examples will suffice. In Victorian times small fishing villages were much praised as models of traditional morally upright community in contrast to the evils of the dark, dirty, industrial city. In many ways they were anachronistic, and their key personnel consisted of aged fishermen and fishwives. The memory of such idylls has remained strong in the British psyche and, heritage aside, helps to explain why, along with the health-giving benefits of salt water and sea air, many middle-class retirees have left the city suburbs to live in seaside towns. Observers have noted that such towns are 'slumless, well-ordered, obsessively neat. Crime and vandalism are rare, social problems hard to find. There are no down-and-outs, no drunks, no scroungers; apparently no poverty' (Seabrook and Roberts, 1981: 26). In short, such towns, dotted around the British coast, stand as a model for traditional community values. They perform a symbolic role as a critique of modern society and only manage this because from the time of their establishment they have been well out of the industrial mainstream and thus cushioned from its worst effects. But whatever the seeming marginality of such places they are very real – by 1951 some resorts had populations comprising one-third of people over 60. This pattern has been consolidated as more people expect to live longer, with the incentive to plan their retirement in relative economic independence from younger family. The consequent picture of large swathes of the coastline of southern England is thus one of a Costa Geriatrica. Arguably this situation shows how older people have become agents in the reconstruction of communities in their own desired image (Blaikie, 1997).

Such a pattern suggests the triumph of conservatism. However, Holdsworth and Laws, examining coastal British Columbia, talk of the redevelopment of the 'land of old age', the ways in which new built environments of ageing match shifts to a postmodern life course. Part of the shift towards new lifestyles in old age is a corresponding cultivation of designer landscapes, be these bungalows, or congregate-care complexes, or 'golf-and-country-club estates with secured perimeters'. The development of specialised housing sub-markets has heightened a tension between 'a built environment still dominated by a legacy of the modernist preoccupation with order and segregation and a postmodern society which is increasingly moving away from a strict model of chronological age scheduling'. Under modernity, each life course stage was dominated by a particular built environment: school, workplace, home and poor house/institution. In the postmodern city, property developers have exploited the discretionary wealth of some older people to redesign . . . 'to refit suburbia for lives of leisure but lives of gradual frailty, as distinct from the lives of child-rearing and commuting for which they were first built'. Against this, conventional old people's homes house a relic population of frail fourth-agers who are thus 'largely excluded from the new construction of the life course' (Holdsworth and Laws, 1994: 174, 176, 179, 181).

Having control over how and where time is spent is a key aspect of leisure, and much of the third-age literature adopts a liberationist perspective whereby older people freely experiment with new ways of living. Del Webb's Sun City West advertises: 'A friendly hometown. With rolling greens and community spirit' in its 'planned small town atmosphere' (ITV, 1997). Arguably these are not communities at all, but what Bellah *et al.* call lifestyle enclaves, networks:

> formed by people who share some feature of private life. Members of a lifestyle enclave express their identity through shared patterns of appearance, consumption, and leisure activities, which often serve to differentiate them sharply from those with other lifestyles. They are not interdependent, do not act together politically, and do not share a history.
>
> (Bellah *et al.*, 1988: 335)

Thus 'whereas a community attempts to be an inclusive whole, celebrating the interdependence of public and private life and of the different callings of all, lifestyle is fundamentally segmental and celebrates the narcissism of similarity' (p. 72). Certainly, such age-segregated living arrangements as Sun Cities, retirement villages and seaside towns capture one of the great ironies of post-modernity in that an enhanced range of choices renders desirable the option of living in a group that is fundamentally characterised by its social and cultural homogeneity. Similarly, Longino and Marshall's study of Canadian 'snowbirds' migrating to Florida for the winter revealed that they were 'nomadic in the sense that their social ties were primarily with the same migrants in the communities they shared at both ends of the move. Their ties were not to places but to the migrating community itself' (Longino and Marshall, 1990: 234).

We might therefore consider the lifestyle enclave an appropriate form of collective support in an otherwise radically individualising society. The segregationist impulse is one of 'encrustation', a barricading-in which has 'become increasingly common as senior adults find themselves in a world that does not seem to understand and respect the values they so long cherished' (Kastenbaum, 1993: 180). To this extent elders are not so much projecting their desired lifestyles onto the spatial template as using space to withdraw from the mainstream. Retirement thus becomes a form of privileged retreatism, fully in line with the leisure images retailed in the advertising. Meanwhile, the opportunity to relocate remains the privilege of a minority. For most pensioners, maintaining a sense of identity as they ride the doldrums of deep old age is a matter of clinging on to lifelong convictions and carving out acceptable lives in the context of an overarching dominant culture which devalues ageing.

Thus far we have only considered the minority, the vanguard groups who trailblaze a different physical environment in which to age dynamically. But most of our elders live rather more mundane, less appealing lives. They do not live out their dreams. Without the visible articulation of their values in physical form how might we understand them? One means of comprehending attitudes in later life is by doing ethnographies that foreground personal narratives. Such narratives are important, not only because they demonstrate the rationales behind individuals' ways of interpreting their lives, but also because they provide insight into the cultural bases on which particular understandings have been developed. These may be linked to vocabularies of motive, the specific ways in which people justify their views and behaviour (Mills, 1940). For instance, Williams has considered the importance of the Protestant work ethic in directing the perceptions of older Scots about death and illness, his argument being that a generational world view may act as a categorical imperative by which all questions of morality are tested (Williams, 1990). His subjects regarded themselves as duty-bound to remain active and resist the onset of ill health. Only when serious illness had been medically diagnosed did they consider disengagement legitimately sanctioned. For them, illness hinted at sloth and sin while good health reflected virtue. By extension, therefore, the continuing commitment of today's over-75s to the work ethic, thrift and orderly routines is a legacy of their lifelong beliefs which are grounded in clear moral categories. The extent to which these underpinnings are necessary is borne out in the critique of modernity, for in the contradictions of contemporary individualism lie the roots of a quest for moral community.

Involution – looking backward?

In *Modernity and Self-Identity*, Anthony Giddens discusses what he calls tribulations of the self, anxieties with which the individual must cope (Giddens, 1991). Risk and doubt produce worries that may be countered by religious faith and /or opting for a lifestyle that reduces these or, at least, means the individual does not have to think about them, hence – to a degree – the moral certainties of Williams' *Protestant Legacy*. But more significantly for our purposes, he talks about ontological security – the sense of safety of one's being; that is, the maintenance of identity and the self. Here globalising tendencies present a major threat since they are bound up with losing a sense of place and identity. Sociologists argue that postmodernity, consumer culture and other contemporary developments each affect lifestyle choices and changes that are beyond our control, thus leaving us with the sense that our lives are being taken out of our own hands. One consequence is the withdrawal of individuals from society in order to preserve their ontological security. If these tribulations affect us all, then how much more must they affect older people whose vulnerability and powerlessness increases in direct proportion with their anxieties?

We can extend this notion of preservation through recreation of past worlds by considering the idea of imagined community. If we can talk of the land of old age, it is perhaps appropriate that we consider how elders develop an allegiance (or otherwise) to that country. Anderson argues that nationalism requires people to regard the nation as an imagined community. It is imagined in that it requires a communion with people one has never met. There must be a deep sense of comradeship with others through feelings of belonging rooted in both time and place. Historical continuity is vital, while the notion of community implies a boundary separating insiders from outsiders. Finally, for the community to be a true nation there must be self-determination (Anderson, 1993). It would be stretching the analogy too far to suggest that older people – or, at least, most of them – would wish to rise up as a nation of the old. Nevertheless, when used to investigate how people visualise their ideal moral universe, the idea of imagined community does bear consideration as a heuristic device through which world views may be articulated. Recent research indicates that older people who feel socially marginalised try to make sense of the world using notions of imagined community. As the social circle narrows with the death of spouse and friends, some seek solace in identification with the past-situated mores of dead relatives. They evoke better times when families and neighbours were 'a little world on our own', then cast this idealised image onto their values in the present. Such fabricated re-connection with the past creates a mental and moral resource for sustaining biographical continuity while helping individuals to cope with a fractured and bewildering present (Conway, 2000). Similarly, in a study of women moving to old-age facilities, Redfoot found that middle-class women, having established strong attachments to place, often coped with relocation by making their new rooms into museums of their lives. Thus they remained committed to their former existence, and to the ontological security this provided, by conserving its material culture, their accumulated lumber of a lifetime (Redfoot, 1987). With such household goods go the values attached to them. The desire of old people for a consistent identity bridging past and present thus resembles the quest for national identity in that both involve a struggle of memory against forgetting.

In these settings, since any community is imagined rather than real it only serves as a consolatory or compensatory backdrop to a stark and dreary present; it is fundamentally inward-looking. It seems, however, that the media are wise to the requirements of their elderly audience, since through such popular media as television older people confined to their homes sometimes enter a singularly private and sheltered realm where television programmes themselves may suffice in delivering their daily dose. *Coronation Street* is Britain's longest-running and most popular television soap opera. Its moral community has been dubbed a 'fictional celebration of a kind of world we have lost: a nostalgic recreating, larger than life, of the enduring values of life on a small and intimate scale where survival is based on interpersonal accommodation and respect. Such values [are] reflected in the pride of place occupied by the older characters' ('Free Radicals', 1997: 25). For those watching – and there are millions – the community is 'imagined' in the sense that actual interaction with the TV actors is impossible, yet it remains a 'community' because, regardless of any actual degree of inequality, its relationships are always those of 'deep horizontal comradeship', and its values those with which older viewers identify. Likewise with the almost equally popular soap serial *EastEnders* in which the values of the urban village are constantly being addressed and where older characters play significant roles as community watchdogs.

The cover of the weekly magazine *The People's Friend* nearly always depicts a country scene. Look inside, and among the ubiquitous love stories, one reads regular features called 'The Farmer and his Wife' and 'From the Manse Window'. Contents analysis reveals interviews about the filming of romantic TV serials: in one an actor talks of how one of these highlights the disappearing values of neighbourliness. There are holiday features on Balmoral, the Welsh hills, the Yorkshire Dales, the Lake District, and poems about 'Highland Colours' accompanied by pictures of Glencoe. Images of city streets are rare indeed. The fact that stairlifts and incontinence aids predominate among the advertisements is sufficient reminder, however, that 'The Famous Story Paper for Women', as it styles itself, is marketed primarily at old ladies. One might ask why essentially rural romance should be so readily associated with such a social group. A plausible answer would be that, from the vantage point of today's elderly reader stuck in her inner city tower block apartment, these images could trigger comfort in a world of urban anomie bereft of the mores of community. The magazine purveys an alternative vision where, instead of vandalism, theft, muggings and disorder, justice and harmony prevail and everyone lives happily ever after. Not only is its signature nostalgic and escapist, but *The People's Friend* also provides devices, such as the readers' letters page, through which readers interactively construct the moral vision.

Such idealisations can, of course, be shattered when rudely confronted with a return to the real, although fiction itself has generated sufficient counter-mythologies. Novelist Alan Sillitoe reveals these in *Birthday*, his sequel to *Saturday Night and Sunday Morning*, written over forty years previously. Like the writer, the protagonist has also aged forty years. During this time we are told that Brian Seaton has lived and worked in London, but suddenly he revisits the provincial town he grew up in to see his brother and attend his childhood sweetheart's 70th birthday. The two brothers roam around the area, visiting old haunts – 'no matter how changed, it was an area in which he had no need of maps. Everything was in the past, but an event could leap to mind with such intensity it might have happened in the last five minutes' – yet Seaton finds himself facing disillusion and disappointment, the point being that his mental image of this fictional Nottinghamshire is one anchored firmly in a reality that has long vanished (Sillitoe, 2002).

Similarly, the recent film *About Schmidt* transposes the road movie plot, in which the physical journey across America is used as a metaphorical space for psychological dislocation and travel, from youth to later life. Schmidt is 'a man at a crossroads' moving from 'a productive working life into a bleak retirement . . . he escapes the life that has formed itself around him' by taking to the wheel of a Winnebago motor home.[2] An uptight and 'repressed conformist undergoing experiences he doesn't wholly comprehend', Schmidt is 'a man with an unhappy inner self, but no supportive inner life other than that provided by the conventions and values of middle-class middle America' (French, 2003: 7). The narrative consists of a sentimental journey to the town where he was born, to the campus of his alma mater, finally to the home of his daughter and her imminent in-laws. At every stage he finds despair and deep alienation. The Winnebago merely underlines his loneliness. His only bond becomes an imaginary one with an African boy whom he 'sponsors' by sending money and writing occasional letters. Notwithstanding the excruciating mismatch revealed in his 'blinkered expectations of what a child in an East African village might grasp about American life', he finds redemption in what he perceives to be his sole cross-generational identification in an otherwise worthless existence (French, 2003: 7). Unlike Sillitoe's Seaton, Schmidt seeks consolation in an imaginary connection outside the bounds of the altered reality round about him. Yet the message is, if anything, still bleaker; we find pathos in his desperate, misplaced awareness. Ultimately, there is no escape from the lifetime that has made him who he is. (As French indicates, his despair at the end of the film connotes a very different message to that of Kurosawa's *Ikuru*, where the elderly civil servant finds fulfilment in creating a small children's park (French, 2003: 7)).

In each of the above, the challenge of maintaining a coherent identity has been enforced either by the relocation of the individual or by change in the places to which they cleave. Under conditions of high or late modernity, the fixity of people and place, of permanent communities, is rare indeed. Nevertheless, among the heroes ageing are a few stalwart survivors whose contentment (and longevity perhaps) owe much to their tenacity. Elizabeth Cockburn lived without electricity or telephone for 77 years in the same remote cottage in rural Aberdeenshire. She only removed, aged 100, after being cut off by the snow for three weeks in the winter of 2000–1. Widowed some fifty years ago, she worked single-handedly on her 56 acres of land without any of the conveniences of modern living. Her peasant-like attachment to locale without modern domestic conveniences, akin to the celebrated autobiography of Hannah Hauxwell who lived alone in a remote Yorkshire farmhouse, with neither water nor electricity, is something of a cause for wonderment in this day and age (Simpson, 2002: 10; Hauxwell, 1989).

Conclusion: mirroring ourselves

Our brief excursion has taken us to some exotic places, and some rather more mundane, both in the physical world and in the mind. Yet we should recognise that not only are the landscapes of later life altering as contemporary third-agers come to terms with the possibilities and choices of active retirement while the less fortunate fourth-agers eke out an existence from a more barren land, but the time-travellers themselves are also changing.

Bauman argues that in the past identity was built up, layer-on-layer, from experience. Our modernist forebears were 'pilgrims through life', guided by a 'life project' and belief in progress. However, today's identity-seekers are simply nomads who 'hardly ever reach in their imagination beyond the next caravan-site' (Bauman, 1992: 167). Destiny has been supplanted by uncertainty. For Bauman, 'the overall result is the *fragmentation* of time into *episodes*'. Instead of building coherent life strategies through which a sense of purpose emerges, we shift as we go along. He characterises the change as being from a preoccupation with the 'daunting task of identity-building' to a sense of being 'moved by the horror of being bound and fixed'. As Bauman underlines, 'the foremost strategy of life as pilgrimage, of life as identity-building, was "saving for the future", but saving for the future made sense as a strategy only in so far as one could be sure that the future would reward the savings with interest' (Bauman, 1996: 23). With the fast-eroding value of the state pension many potential elders may dispense with notions of thrift. Meanwhile, against the quest for community evident in, say, retiring to the seaside, there exist countervailing desires for more individualistic pursuits: the 'performing self' obsessed with surface appearances (facelifts and makeovers), the flâneur whose domain is the shopping mall and the Internet, the vagabond who decides where to turn when he comes to the crossroads (the three million-plus Americans over age 55 who have taken to the road in their recreational vehicles, known as RVers), and the tourist beloved of the Saga holidays. Bauman concludes, 'the real problem is not how to build identity, but how to preserve it' in a world where identities can be adopted and discarded, are disposable, instantly obsolete.

How, then, do we frame our own identities as we age? How do we place ourselves? What utility do symbols have for the creation and maintenance of coherent identities? Or do they reflect the dispersal of identities as theorised by Bauman? Evidence from the research – Kaufmann's aptly titled *The Ageless Self* being a good example – suggests that because we regard identity as central to understanding ourselves, we will go to considerable lengths to sustain a coherent sense of self. Yet, at the same time, Beck contends that the attenuated development of industrial society has led to 'reflexive modernisation', whereby individuals become increasingly detached from ties of family, class and locality (Beck, 1992). As a consequence, we fashion our own choices, identities and biographies and are related to others only through the market. Arguably, individual lives have become personal 'projects' revolving around lifestyles while the notion of an integral, coherent self has been fragmented so that identity can change at different stages in the life course. There is thus a tension between the biographical idea of a continuous developing self and the postmodern notion of the multiple, situational self. For every phenomenological argument that stresses, say, the meaning of 'home' as primitively fundamental to our sense of oneness with the world, there is a postmodern counterargument that posits well-being as an evanescent goal attainable only by constantly adapting to cultural revaluations (Bachelard, 1994; Featherstone, 1991). Armed with the right road map, we are able to travel light, unencumbered by the domestic cultural baggage that so weighed down our forebears, but do we remain content 'in ourselves' as we age?

In evaluating differing perceptions of place among our elders, we are not looking, however, at a tension between expressive, other-wordly desires for Utopia against instrumental, real-world struggles for a better place in the consumerist land market, so much as at how differently placed individuals and groups operationalise their imaginings. For the most part, where people find themselves in later life is not where they would wish to be. Perhaps thinking about metaphors of place in general, and imagined communities in particular, can help us to see how individuals deal with this difficulty. We all construct our own microcosm of the world. At crucial times in our lives – retirement from work or when a wife becomes a widow – the loss of one's sense of historical understanding of oneself may provoke identity crisis while the absence of a sense of present identity provokes existential anxieties. Less explicitly, our sense of self undergoes constant re-evaluation. Periodically, a new status has to be negotiated at an intermediate level between that of the broad cultural scale and that of the individual personality. That intermediate world is constituted by the social network where our identities are created, confirmed, maintained and changed through interaction with other people. Therefore, the self is reflexively reconstructed through formal or informal social contacts within a limited community. But we also create imagined worlds that we inhabit just as fully.

At heart, understanding ageing is itself an imaginative project and the role of metaphor is central. In doing our imagining, the ability to transpose time and space is juxtaposed with a readiness to substitute people for place. Just as the ageing body can be likened to ageing scenery, so the world itself may be lent human characteristics. Jan Morris explores this idea in her last book, *Trieste and the Meaning of Nowhere* (2001), when she says, 'The allure of lost consequence and faded power is seducing me, the passing of time, the passing of friends, the scrapping of great ships!' This time, however, the metaphor is reversed, for it is places, and Trieste in particular, that are likened to people, rather than the other way about. She continues anthropomorphically: 'a great city that has lost its purpose is like a specialist in retirement. He potters around the house. He tinkers with this hobby or that. He reads a little . . . does a bit of gardening. But he knows that the real energy of his life . . . is never going to be resumed' (Morris, 2001: 2, 153). Likewise, ageing bodies are likened to archaeological ruins, old buildings to rotting teeth, 'rocks as bones, soil as flesh, grass as hair, tides as the pulse', nations as young or mature, or decadent. All are alike in the repugnance evoked by the 'look of age'. Polybius compares countries to people and talks of nations being 'vaunted in youth and belittled in old age' (Lowenthal, 1985: 128, 140). Decay is an inescapable consequence of youthful flowering, and is seldom admired. Equally, however, the look of age can be the signal of antiquity, which although it needs decay to guarantee its authenticity, is nevertheless much prized. Wrinkles might therefore inspire awe and suggest wisdom even while inciting physical revulsion. What all this points to is a fundamental reflexivity between the self and the surrounding material landscape. Indeed, Morris gives the last word to Jorge Luis Borges, whose artist (like Morris herself) set out to portray the world, but discovered that his 'patient labyrinth of lines framed his own face' (Morris, 2001: 185–6).

Most of us, the un-aged, are familiar with older people, and we might imagine ourselves being old, but what we are doing is visiting the country as temporary tourists; we are not ourselves part of the landscape. Ironically, our distance from it grants us a perspective often denied those whose very being in the place renders it so familiar as to appear commonsensical. Yet we find it difficult to grasp how that common sense differs from our own. Neither subject nor observer is therefore well-placed to comprehend *both* what it is that makes the context of later life distinctive *and* what it feels like to live there. Arguably, though, using metaphor helps us to bridge the gap.

Notes

1 This paper develops ideas originally advanced at a University of Bergen/European Behavioural Social Science Research Section (IAG) symposium on 'Ageing in Europe: Images and Identities' held in August 2002. I should like to thank fellow-participants for their comments.

2 Quotations from www.picturehouses.co.uk programme notes (2003). As the later-life owner of a recreational vehicle, Schmidt appears to be part of a growing trend across North America. However, both in his reluctance to take to the road and his motivation and behaviour he remains unusual (cf. Blaikie, 1999: 178–9).

13

CATALOGUING OLD AGE

Bill Bytheway and Julia Johnson

Old age is a concept that lends itself to visual display; in many different books, journals, magazines and exhibitions, it is represented by photographic images that include older people. The words 'old age' are often used in reference to a series of such images, or perhaps are included in its title. Sometimes the people photographed are famous and named, but often they are not and then the series resembles a researcher's sample: the implicit intention is to generalise about old age. The viewer, like the researcher, is able to move from image to image, both reflecting on a slowly accumulating, aggregated sense of what old age looks like and, at the same time, appreciating and analysing the diversity of the presented images.

A series of photographs of old age has to be compiled. This may be undertaken by a publishing editor, a designer, an exhibition organiser or a conference committee (or, indeed, the photographer). Photographs may be selected from the work of one photographer, from many or from photo libraries, either in-house or commercial. A series may be presented for public display only when it is fully compiled, or selected images may appear periodically, on the cover of a magazine, for example. The aim of the compiler may be purely documentary: to illustrate old age as currently experienced by older people perhaps. Or the aim may be to select images that convey a message. The message may be negative – later life is loss and deprivation, for example – or it may be positive – later life can be fulfilling – or it may be more complex. Whatever, the compiler sets out to select photographs that will meet the particular aim for the series. Each photograph may be supplemented by a caption which reinforces an intended interpretation, and the whole series may be accompanied by a written text: a book, catalogue, brochure or whatever.

Well-known examples of series of photographs

The 1978 exhibition *Images of Old Age in America, 1790 to the present* is a famous example (Achenbaum and Kusnerz, 1978; Blaikie, 1999: 125–6). Like projects such as *The Family of Man* (Steichen, 1955), *Images of Old Age in America* was intended to contribute to the development of a collective American consciousness by documenting the interplay between human lives and historical circumstance (Stott, 1973; Sontag, 1977). Approximately 100 images were included and it toured the US – colleges and banks as well as art galleries – and was displayed at the historic 1981 White House Conference on Aging. The objective was 'to explore how the cultural-historical factors in the US had shaped *the meanings and experiences of old age*' (Achenbaum, 1995: 20; emphasis added).

Another example is the cover of *The Gerontologist*, the journal of the Gerontological Society of America. Routinely this includes a single photographic image of later life. Most often this is a portrait of one particular person. Over the years, the objective of the editors of the journal appears to have been to present diverse and positive images of being old. For example, three recent editions include photographs with the following captions:

- 'Donna Day's photograph of four women stretching in a lake portrays the spirit of successful ageing' (edition 43, i)
- 'Willie Mae Brown, 70 years old, is a resident of Tandem Health Care of Tallahassee' (edition 43, ii)
- 'The Franciscan Handmaids of Mary are a mostly African American order of nuns in Harlem. This photo was taken during their 85th anniversary celebration in October 2001' (the caption accompanying the portrait of one unnamed nun, edition 43, ii).

It is interesting to note that there is no set pattern to the wording of these captions. Sometimes the age of those being photographed is made explicit (as in edition 43, ii) but sometimes not (as in the other two). Similarly some captions, such as that of edition 43, i, include explicit messages about how to age or how to be old, whereas others do not. Often the location is named (as in 43, ii and 43, ii), but sometimes not.

Many books relating to age similarly contain illustrative photographs, often including one on the front cover. Well-known examples are Elder (1977), Myerhoff (1980), Macdonald and Rich (1984), Silverman (1987) and Thompson *et al.* (1990), and, more recently, Featherstone and Wernick (1995), Means and Smith (1998), Blaikie (1999) and Gubrium and Holstein (2000). Some of these photographs were chosen from photo libraries. Publishers have always appreciated how striking photographs help to market books.

This chapter reports an analysis of material that is part of an ongoing study of images of age and ageism. In Johnson and Bytheway (1997) we studied photographs appearing over a 16-year period in a UK social care magazine. In Bytheway and Johnson (1998) we turned to a variety of sources including material from a course in cartoon drawing. Bytheway (2003) compares and contrasts a volume of 82 photographs by the celebrated Californian photographer Imogen Cunningham with images appearing in a popular UK magazine aimed at older people. The focus in all these analyses has been on the interplay between image, words and age. We have drawn upon a growing literature on the analysis of visual images (Hevey, 1992; Emmison and Smith, 2000; Rose, 2001).

Picture libraries

The aim of this analysis is to build upon that of Johnson and Bytheway (1997). For that paper, we analysed 270 photographs published between 1978 and 1994 in a British professional magazine aimed at social workers and others employed in community care. Each photograph included at least one older person and a younger carer, and we identified four types of portrayed relationship:

- 'teacher-like' images in which the carer is advising or supervising the older person/s;
- family-like portraits;
- images of caring for older people by undertaking domestic tasks; and
- images of 'caring about' older people, often involving touch.

The analysis revealed much about dominant beliefs regarding good practice in caring for older people. Many of the photographs illustrated articles promoting these beliefs. More generally, the objective of the journal appeared to be the generation of a positive image of the care relationship.

A number of the photographs published in the magazine were obtained from the John Birdsall Social Issues Photo (JBSIP) Library. It is a well-established member of the British Association of Picture Libraries and Agencies (BAPLA). In 2003 BAPLA had over 400 member companies, offering, according to its website, 350 million images.

The homepage of the JBSIP Library describes it as 'the leading photo library for images of social issues in the UK'. Its catalogue has over 6000 images, each available for inspection online. The homepage includes examples of how they have been used to illustrate annual reports, posters, websites, brochures, magazines, books and advertisements. Among the clients listed on its website are a number of national organisations and publishers. The Open University too has made good use of the Library in the production of its learning materials. The Library is primarily targeted at agencies involved in 'social issues' and, to this end, it offers images which are relevant to policy, management and practice in health and social care. This is reflected in the thirteen categories into which the Library's catalogue is organised. The most numerous categories, each with over 500 images in August 2003, include children, disability, health, families, education, housing and homelessness, and young people. Under the heading of 'old age' there are 320 images.

Each photograph is accompanied by a code number (indicating which photographs were taken on the same film), a two-part caption (e.g. the first is 'Elderly couple dancing – Elderly couple dancing and singing at day centre') and a number of keywords. Some of these keywords are in German (reflecting the Library's success is marketing its pictures in Germany) and some relate to technical details. The remainder, however, provide direct evidence of what the Library perceives to be the dominant characteristics of the image (e.g. 'smiling', 'anxious', 'garden') and how the image might be used (e.g. 'tobacco', 'isolated').

Who is included in photographs of old age?

As with *The Gerontologist*, the captions that accompany the photographs provide information regarding content and interpretation. All 320 captions identify the subject of the photograph as a person or persons. The word 'elderly' is used in 87% of the captions. About the remaining 13%, there is little that is distinctive and, in particular, it is not difficult to identify 'elderly people' in the photographs. Only two photographs do not show a face or head: one of an alarm button hung round the neck and another of a woman's hands delving into a purse. Overall, therefore, old age is represented primarily by images of 'elderly people'. The JBSIP Library, like *The Gerontologist*, places an emphasis on representing cultural diversity and there are 27 elderly people described as black or Asian.

There is reference to one other person in 95 of the captions (30%) but only once to a third ('Woman and child with elderly man in day centre'). For 77 of these (81%), this person is described as 'a carer'. For example, the sixth photograph in the catalogue (the first to feature a carer) has the caption 'Carer helping elderly woman with stick to walk'. That so many photographs include 'a carer' is powerful evidence of how, in social issues literature, old age has come to be associated with care, and the 'elderly person' with 'a carer' (Johnson and Bytheway, 1997; Johnson, 1998). In the wording of 71 of these 77 captions older person and carer are identified jointly as the subject of the photograph (e.g. 'elderly man and carer', as opposed to 'elderly man with carer'). This indicates that in these images of old age the carer is as central as the older person.

Table 13.1 gives the basic statistics about how persons are detailed in the captions. The categories distinguish

- the sex and number of older person(s),
- whether they are described as 'elderly',
- the presence of other persons and
- whether the other person is described as a carer.

These statistics demonstrate that nearly half of the captions refer to an 'elderly woman' and a quarter to an 'elderly man'. Perhaps the most striking aspect of the table is that 46% of the photographs of an elderly woman include 'a carer', compared with 9% of those of an elderly man and 27% overall. As many as 59, or 18% of the whole collection, are of an 'elderly woman' and a female 'carer'. This is evidence of how the image of care in old age is gendered, both in terms of provider and recipient.

Table 13.1 How persons are detailed in captions: some basic statistics

Older person(s)	*Female carer*	*Male carer*	*Other female*	*Other male*	*No other person*	*Total*
Elderly woman	59	4	7	1	78	149
Elderly women	3	0	1	1	14	19
Elderly man	5	1	4	0	64	74
Elderly men	0	0	0	0	4	4
Elderly couple	0	0	0	0	24	24
Elderly people, etc.	0	0	0	0	10	10
Woman	2	1	1	0	15	19
Women	0	0	0	0	2	2
Man	1	1	1	0	6	9
Couple	0	0	0	0	3	3
People, etc.[1]	0	0	1	0	6	7
Total	70	7	15	2	226	320

Note

1 'People, etc.' includes terms such as 'residents' and 'pensioners'.

Where are the photographs taken?

There are 78 photographs that are described as portraits. Most of these conform to the usual composition of a studio portrait, providing virtually no information regarding location. A few appear to be taken outside, in a garden or against a wall, but again the intention is clearly to photograph the face and shoulders of the subject and little else. Setting these to one side, among the remaining 242 images, there are 176 indoor locations (73%). This indicates that the dominant image of old age is located indoors.

Of these 176 indoor photographs, there are 73 where there is no wording, either in caption or keywords, indicating a specific location. Of the remaining 103, there are 63 (61%) where there is a direct reference to a residential or care home or to a day or care centre. Further, only 15 are specified as having been taken in the older person's own home. Moreover, there are a number of other indoor locations that are indicated by captions or keywords: hospital (3 photographs), pub (2), library (3), exercise class (2), supermarket (3), shop (1) or swimming pool (2).

This suggests that care settings are the primary indoor location. Certainly, the captions and keywording direct the viewer's attention to images of old age in care settings. That said, however, there are clues in the photographs themselves (and in their Library code numbers) to indicate where they were taken. For example, fire escape notices indicate that it is a care setting rather than the subject's own home. More commonly, the furniture and seating arrangements help to distinguish between formal and domestic locations. On the basis of these judgements, the numbers are adjusted as follows: 80 care settings (45%), 59 own home (34%), 17 other named settings, and 20 unknown. This adjustment indicates that rather more were taken in the older person's own home than the captions and keywords suggest. Nevertheless, care settings are still the most common.

There are 104 non-portrait images taken indoors but not in care settings. It is significant that they include 42 for which the captions refer to a carer. Typically these are images of a carer visiting the elderly person at home and providing various kinds of service. In total, 116 (66%) of the captions of the 176 indoor photographs refer to residential or day care, or to carers. In this way, the images and their captions consolidate the association between old age and care provision.

The fifteen photographs described as being taken in 'own home' differ from those of residential homes and day centres only in the absence of groups of older people and notices. Where they include carers they all include symbols of care such as the carer's uniform. Despite this the inclusion of the word 'own' in the caption promotes an interpretation of the image in terms of privacy and territory and, as a result, the various furnishings that are visible can be seen by the viewer as belonging to the older person. 'Carer listening to elderly woman play the piano in own home', for example, is a caption that conveys a particularly distinctive interpretation of care and home. However, it is rare for the carer to be presented in such an inactive state as 'listening'. Several other photographs are similarly domestic and seemingly private, being located 'in kitchen' or 'in bedroom'. Nevertheless it is important to note that these images of older people in their 'own' settings are comparatively few in comparison with communal care settings.

In total there are 66 non-portrait photographs that are taken outdoors. There are 14 that are explicitly located 'in garden', 'in allotment' or 'in greenhouse' and a further two captions refer to elderly people 'gardening'. All these photographs convey a sense of activity, pleasure and achievement. There are a further six photographs which appear to be taken in the garden of a residential home. Taken together this is evidence that the garden is a location of some significance in representing a positive side of old age. Other outdoor photographs tend to reflect other familiar images of 'successful ageing' – playing bowls, for example – and there are six photographs of older people out walking in the countryside or along cliff tops. None of these photographs were taken in poor weather.

Old age 'in the community' is represented by a number of images of women or couples shopping, of men in pubs and of individuals in libraries. The JBSIP Library has a clear policy on the use of 'models' and it may be that the complication of obtaining the permission of those being photographed has inhibited the inclusion of images of people exchanging money for purchases. So the interior of shops is represented by three images of a man pushing a trolley past supermarket shelves and another of a wheelchair-user choosing a handbag (seemingly with no assistance). There is also a photograph of a couple looking at produce on an outdoor vegetable stall. Otherwise shopping is represented by people carrying bags along wet streets, and by women laden with purchases climbing onto buses. There are at least four photographs which are less specific but nevertheless set in outdoor urban settings indicated by roads, pavements and brick walls. One notable caption, the only one that includes reference to someone other than a health or social care worker, family or friend, is 'Elderly man on estate talking to community policewoman'.

What are people doing in these images of old age?

There are two distinctive kinds of caption, reflecting contrasting types of photograph. A total of 150 (47%) interpret the image as an action, or a state of being, on the part of the older person. So what is photographed is 'the elderly woman reading' rather than 'the elderly woman'. The second type is the portrait: a total of 78 captions (24%) include the word 'portrait' and here the comparatively few captions containing words that indicate actions or states of being tend to be passive, such as 'looking sad'. So these are photographs of the person rather than an activity. There remain 92 photographs including such images as the actions of carers (e.g. 'Carer serving elderly man his meal at day centre').

Most of the images based in care settings involve actions and convey messages about older people being happy and productive. This is reflected in captions such as

- 'Elderly couple dancing and singing at day centre',
- 'Elderly residents cooking' and
- 'Two elderly women reading and knitting in a residential home'.

Notably there are three photographs of one particular 'elderly man' who is 'volunteering' by cooking and serving food in a day centre. Other activities include playing dominoes, bingo or cards, or eating and talking. Many of these photographs portray older people in groups, sometimes with carers participating too. The keyword 'active' or 'activity' is attached to a number of these images. The participants are sitting in armchairs in a lounge, or sitting around a table, and in the background there are pictures and notices. In contrast to these images of collective activity, a few are more ambivalent and passive; for example, one caption reads, 'Elderly woman looking out of window at residential home'.

There are relatively few heroic images of a super-active later life, such as the skydivers and rock climbers who occasionally appear in advertisements. Three images that appear to be included in order to challenge stereotypes are of a couple in motorbike leathers, a man completing a marathon and a man with a movie projector. Representing 'ideal' examples of 'keeping fit' there are two photographs of women in a swimming pool and two of aerobics classes. Similarly, representing laudable political action, there are two photographs of banner-waving protests, regarding income and housing respectively.

Positive images greatly outnumber negative ones. This is evident in that 47 have the keyword 'happy' compared with only 11 for 'unhappy'. There are 52 subjects who are 'smiling' and 20 who are 'laughing', compared with twelve who are 'sad' and nine who are 'anxious'. Similarly, there are 21 images of 'independence' and only one of 'dependence', and 20 of 'health' compared with none of 'illness', 'ill health' or 'disease'. Regarding the content of the photographs, there are a number that are ambiguous. For example, there are ten photographs of people in wheelchairs and four of people using walking frames. In most of these images, a carer is involved in facilitating mobility and the dominant motif is that of dependence. In two of these, however, the older person appears to be responding positively to the mobility offered by the wheelchair; for example, the woman choosing a handbag. Less ambiguously, there are six images that illustrate fuel poverty and the struggle to keep warm. They all feature the same woman and may have been taken for one particular commission. This image of misery is matched by two other photographs of an older woman crying. One of the captions reads, 'Elderly woman covering face with hands, in despair'.

Ageism and stereotypes

Ever since 1969, when Robert Butler coined the term 'ageism', negative stereotypes of old age have been deplored by gerontologists and by groups campaigning against age discrimination. One concern of the organisers of the *Images of Old Age in America* exhibition, for example, was that it should not 'reinforce' ageist stereotypes (Achenbaum, 1995: 22). The word 'stereotyping' occupies a key position in Butler's well-known definition: 'ageism is a process of systematic stereotyping of and discrimination against people because they are old, just as racism and sexism accomplish this for skin colour and gender' (Butler, 1975: 35).

Featherstone and Hepworth (1990) define negative stereotypes of age as representing a form of symbolic stigmatisation which finds its way through to practical everyday action, thereby giving meaning to the experience of growing old (p. 253).

Note how this definition, like Butler's, implies a process, a chain of actions and reactions, that links the words and images of stereotypes with the policies, practices and interpersonal relations of everyday life. On what basis are stereotypes judged to be stigmatising, negative and deplorable? According to Dyer (1979), stereotypes are distinguished by

- a process of reducing everything about a person to a few traits, and then exaggerating, simplifying and fixing them as permanent characteristics;
- a symbolic frontier between two opposites, us and them; and
- inequalities of power.

In this way, stereotypes are simplistic, divisive and oppressive (Billig, 1987: 184). Stereotyping is the process which associates older people with specific characteristics that set them apart as different from, and unequal to, younger people.

It is not difficult to see similarities with the stereotyping of ethnic minority groups, disabled people and women, and the anti-ageism lobby has not been slow to draw parallels with the fight against racism and sexism, as is evident in Butler's definition quoted above. Nevertheless one weakness of attacking ageism by exposing stereotypes is the implication that 'we' know the truth and 'you' are deluded by false images (this is particularly so when those who are publishing 'the truth' do not count themselves among the old). This is clearly evident in those who attempt to challenge ageism by exposing 'myths' and revealing 'facts' (Rowe and Kahn, 1998). Ageist stereotypes might be more effectively challenged by evidence not that they are false or mythical, but that they do indeed reduce old age to a few exaggerated and simplified traits which reinforce the frontier between young and old.

Featherstone and Hepworth offer the following telling comment on the power of photographs (along with paintings, drawings, sculptures and films):

> The general tangibility and mimetic quality of such representations and the associated mental pictures we carry around with us clearly have an immediacy and facticity which make us think that they are real and self-evident, and ignore the fact that such representations are only able to function within a symbolic order.
>
> (Featherstone and Hepworth, 1990: 252)

So photographic images of old age, they argue, reflect a reality that we can recognise and that we accept unquestioningly in our everyday lives; their qualities lead us to 'think that they are real'. Likewise, many of those working in documentary photography endeavour to produce a comprehensive record of 'the reality'. Photography, for example, has become central to the production of evidence in the prosecution of criminals, and it is often photographs that are perceived to reveal the truth that lead to conviction (Winston, 1998). Similarly, when we see a photograph of a woman with heavy bags struggling to climb onto a bus, we do not doubt that she too would describe it as a struggle. What Featherstone and Hepworth are arguing is that, far from recognising stereotyped images to be false, we are persuaded of their validity by the power of the image. It is as though change can only be achieved by working within the symbolic order of stereotyped images.

So, returning to the compiler of a series of photographs of old age, this argument implies that any intended message will only be accepted as valid if the viewer is able to recognise signs of 'old age' in the selected images. The symbolic order within which old age can be successfully represented includes people in places exhibiting at least some of the following:

- physical signs such as lines on the face,
- age-related clothing,

- aids to daily living such as spectacles and walking sticks,
- an immediate personal space including items such as chair, bed or cutlery,
- an age-related setting such as a residential home or day care centre.

The last of these is important because behind and around any individual in a photograph there is a setting and, as the analysis of the JBSIP Library demonstrates, an age-related setting is often chosen to typify the places in which 'elderly people' live their lives. The furniture and furnishings of day centres and residential homes, the spatial arrangements of seating and tables and the uniforms and positioning of carers are elements in the same symbolic order.

In short, these are the basic elements of the stereotype of old age. Someone who does not have any signs of age such as a lined face, clothing characteristic of older people, spectacles, a walking stick or a carer in attendance, is someone who will look 'far too young' to represent old age. The compiler would reject any such photograph. To be considered valid, a series representing old age can only include photographs that include some combination of lined face, old-fashioned clothing, walking sticks, care setting, etc. It is only when these basic criteria are satisfied that the compiler can begin to select images that support the aim of the series. It is important to note that even if the aim is to demonstrate the diversity of old age (as for example appears to be a prime aim of *The Gerontologist*), the basic criteria still have to be satisfied.

There is a belief which is still widely held that negative stereotypes should be challenged through the publication of positive images (Palmore, 1999: 165). It is commonly assumed that new ways of thinking can be promoted through the use of documentary photographs. In undertaking our study of the care relationship we concluded that the regular publication of images of the 'caring-about' type had contributed to the successful promotion of an image of paid carers who 'care about' as well as 'care for' older people (Johnson and Bytheway, 1997: 137–8). These photographs illustrated articles promoting this type of care relationship. This conclusion is confirmed by the analysis of the JBSIP Library: there is a heavy priority given to positive rather than negative images of care.

Conclusion

The JBSIP Library is a successful commercial enterprise. Like any such undertaking it both leads and follows the market. It is clear from its catalogue categories (and from a casual browsing through contemporary literature) that it is currently servicing a particular niche in the care market. This relates to education and training as well as the development of policies on provision and care practice. Within that world it is selling images of old age. So the concluding question is this: does the evidence from this analysis indicate that picture libraries reinforce stereotyped images of old age, or that they challenge and 'deconstruct' these images?

First it should be noted that there is no indication on the Library's homepage that it aims to promote change. The one issue on which it is keen to lead relates to images of multi-cultural Britain. The second basic point to note is that it includes 'old age' as one of 13 'social issues' categories. It does this without providing any details regarding criteria for inclusion. The fact that virtually all 320 images of old age include at least one person labelled as 'elderly' in the caption confirms the importance of photographs of old age including indisputable images of an identifiable category of older people. The Library chooses to label this category 'elderly' whereas Imogen Cunningham chose 'after ninety' (Cunningham, 1977). The actual label is incidental; what appears to be essential is the implementation of a categorisation.

In addition, this analysis of the captions and photographs of the JBSIP Library illustrates how photographs 'en masse' can promote certain associations. In particular it confirms the close association between old age and care that we have reported in our earlier analysis (Johnson and Bytheway, 1997). Unlike *The Gerontologist*, in which images of care are rare (edition 41.5 was a notable exception: 'Nursing home resident and aide in a skilled nursing facility in San Bernardino County, CA'), the JBSIP Library sustains, if only through force of numbers, the association, particularly that between female carers and older women.

This analysis also indicates how this interpretation of old age – elderly people in receipt of care – is also associated with particular settings. Care is provided by carers, but it is also provided in residential homes and day care centres. Seemingly it is in these settings that old age is most accessible and most clearly made visible.

So, through the selection of images each of which includes at least some of the essential identifiers of old age, the image of old age produced by the Library is indeed dominated by a few exaggerated and simplified traits of older people (Dyer, 1979). First, there are the familiar physical signs of age, clearly displayed in the portrait (as they are on the cover of *The Gerontologist*) but also evident in many action images. These imply that old age is essentially the experience of living with and managing physical changes in later life. Second, there is 'activity': older people happily engaged in games, gardens, buses, shopping and other social activities. What are featured here are many of the symbols of later life: spectacles, walking sticks, cups of tea and warm clothing. Third, there is the need for comfort and care, and the 'fact' that this is made available to them in care settings where the distinctive furnishings and social life help to recreate the sense of a revived active life.

Next in Dyer's criteria, the image of old age has reinforced, particularly through its captions, a distinction, a 'symbolic frontier', between 'elderly persons' and 'carers'. There are only a few images of older persons who 'care' or of carers who are 'elderly'; in the large majority of photographs there is an indisputable age difference between the older 'elderly person' and the younger 'carer'. Finally, inequalities in power are evident in the asymmetric relationship of 'being cared for' and 'being cared about'. The carers all appear to be healthy, hearty and competent people. None appear to be over-stressed, under-paid or abusive. Conversely, with few exceptions, the older persons are portrayed as happy, submissive and in need of guidance and support.

There are exceptions in the Library that might be used to challenge this stereotype – the images of political protest, motorbikes and countryside ramblers, for example – but these represent a very few images. Arguably each of these is itself a stereotype that the library would be expected to cover: a handful of counter-images that serve to consolidate and reinforce the dominant image that is represented by large numbers of photographs. What is perhaps more striking are those images relevant to social issues that are totally missing from the Library's series: older people driving cars, queuing in post offices, babysitting for the family, dealing with government officials, repairing homes, watching television, washing clothes, running voluntary organisations and, perhaps most significantly, caring for and supporting each other.

Early in the twentieth century, the German photographer August Sander endeavoured to compile a 'photographic index' that would catalogue a 'social hierarchy fixed through a series of objective, almost scientific, images of representative figures or types' (Clarke, 1992: 71). Arguably he was a precursor to the picture library of the twenty-first century. Anti-ageism campaigners should be concerned about the growing power of picture libraries and the Internet to service and promote a stereotyped image of an old age that is overwhelmed by care.

14

MAKING SPACE FOR IDENTITY

Sheila M. Peace, Caroline Holland and Leonie Kellaher

Introduction: environment and identity in later life

> [A] person's behavioural and psychological state can be better understood with knowledge of the context in which the person behaves.
>
> (Lawton, 1980: 2)

Is place identity an important component of self-identity? People perceive themselves in several ways. They have 'multiple identities' which range from those imposed by societal stereotypes through to private understandings of self. How we see ourselves may or may not be contextualised in space or place. Yet aspects of self, associated with role identities, may become spatialised – for example, maternal roles may often be associated with a domestic space or leisure roles with public spaces. Both macro and micro environments are recognised domains of quality of life but their importance may vary across the life course and be of greater value to some people than others. Context, reflected in attachment to a place and/or a dwelling, may form and nurture a crucial aspect of well-being.

The work of Rubinstein and Parmelee (1992) continues to provide one of the most comprehensive accounts of identity in relation to the life course:

> Identity is the sense of who one is in the world, distilled from a lifetime of experiences. From the collective perspective, identity consists of the life course as a cultural construct: socially normative and collectively outlined and accepted life-course statuses and transitions. But at the individual level, every person creates for herself a particularised version of the collective life course, a life story, depending upon her specific experiences and the meaning she attaches to them.
>
> (Rubinstein and Parmelee, 1992: 144)

It is this understanding that continues to be developed as we further explore the relationship between environment and identity in later life.

The ecological relationship between person and environment (P–E) is experienced in places and spaces that may be perceived as personal, familial, organisational, territorial, collective, national, global, spiritual and imaginative/symbolic. They can be referred to as places of the mind, places of living, dwelling places, neighbourhood, community, area, nation, place of birth, places of origin, places of death, places of final deposition. This is a very broad definition of surroundings that may encompass the self and which may be more or less easily defined objectively and subjectively.

In this chapter we consider the development and outcome of British research – 'Environment and Identity in Later Life: A cross-setting study', funded by the Economic and Social Research Council as part of the Growing Older Programme from 1999 to 2004. We begin by locating this study within literature concerning environment and identity in later life, then consider the methodological approaches for understanding such relationships, and finally explore how experiences examined in this current British research extend this debate.

Foundations for new thinking

Writers from the multi-disciplinary fields of anthropology, geography, sociology, psychology and architectural design, predominantly North American and western European, have contributed to our thinking and this research. Several important authors have written influentially on space, time and identity (e.g. Soja, 1989; Massey, 1994). Bourdieu (1977) examined the material culture of everyday life to analyse how this is employed in the construction of individual practices and habits. In developing the concept of 'habitus' he concentrates upon the domestic setting and its materialities. However, while time is implicated in the evolution of customs and so on at individual and collective levels, he does not comment upon age-distinctive patterns (Bourdieu, 1977). Others, notably Lefebvre (1991), have argued about the ways in which people – at any age – live and make sense of a changing world and how past, present and future are constructed through spatial and temporal frames. Giddens (1984; 1991), discussing ontological security, recognises context as influential in supporting personal autonomy over everyday routines in a time when the impact of globalisation on activities and behaviours raises levels of risk. Yet here the influence of age is undeveloped (see Phillipson, 2003).

In contrast, gerontologists, who encompass these disciplines, have concentrated on specific aspects of environment and ageing that engage with the material, social and psychological. We read with interest work that varies from British research on the meaning of home (J. Sixsmith, 1986; A. Sixsmith 1990; Gurney and Means, 1993; Hockey, 1999), to discussions of the village community in rural Finland (Koskinen, 2002), to understanding the development of shared living in CoHousing in Denmark and the Netherlands (BiC, 1994; VROM 1997; Brenton, 2001), to the implications of migration within and between countries (United Nations, 2004), to the effect of chronic health conditions on representations of home (Oswald, 1996).

The literature is therefore wide-ranging and yet here we focus on a small number of North American gerontologists/psychologists, anthropologists and geographers whose work has been most influential and considers ecology and ageing, spatiality in later life and attachment to place.

Ecology and ageing

We opened with the words of psychologist M. Powell Lawton. The ecological model of Lawton and Nahemow (1973) introduced the concept of environmental press, where the competence of the individual could be considered alongside the immediate and wider environments in which they live, enabling some judgement of the degree of comfort and performance potential that the individual could experience. At the centre of the model was an adaptation level where people are able to 'tune out the environment' (see Figure 14.1) by achieving a state of balance. They argued that people became aware of environment through environmental press. This suggests a fluidity about environment – as a benign presence, engaging but not threatening. Environmental press might be activated at both macro and micro levels, for example, through neighbourhood change or a change of health status leading to a disabling environment. How an older individual copes with change will relate to personal competence, personality and whether they can adapt behaviour in order to cope and maintain well-being in the face of an environment that has become pressing, if not oppressive, or adapt the environment to facilitate preferred activity.

Lawton and Nahemow (1973) define competence in relation to biological health, sensori-motor functioning, cognitive skill and ego strength – measured through proxy indicators of health and cognitive functioning. This definition of competence does not encompass a sense of growth, rather it is more associated with decline and in earlier work Lawton and Simon (1968) set out the 'environmental docility hypothesis', which stated, 'The less competent the individual, the greater the impact of environmental factors on that individual'. 'Perceived competence' – the way in which people perceive themselves, which is more allied to self-esteem and life quality – was under-developed in earlier work. However, in later research Lawton suggested that the tension felt by environmental press could be seen as positive, as well as negative, in that it caused people to respond and, if possible, to engage with their environment through a form of stimulation and problem-solving (Lawton, 1983; see Weisman and Moore, 2003; Scheidt and Norris-Baker, 2004, for further discussion). The development of this research during the latter part of the twentieth century has been parallelled by other North American research concerning person–environment interaction in later life, addressing issues of need (Kahana and Kahana, 1982; Carp and Carp, 1984; Carp, 1987; Kahana, Kahana and Kercher, 2003) and assessment (Carp, 1994).

Further development of issues concerning personal response to environmental press are also developing through a growing interest in environmental design and the material environment, which has hitherto tended towards a more func-

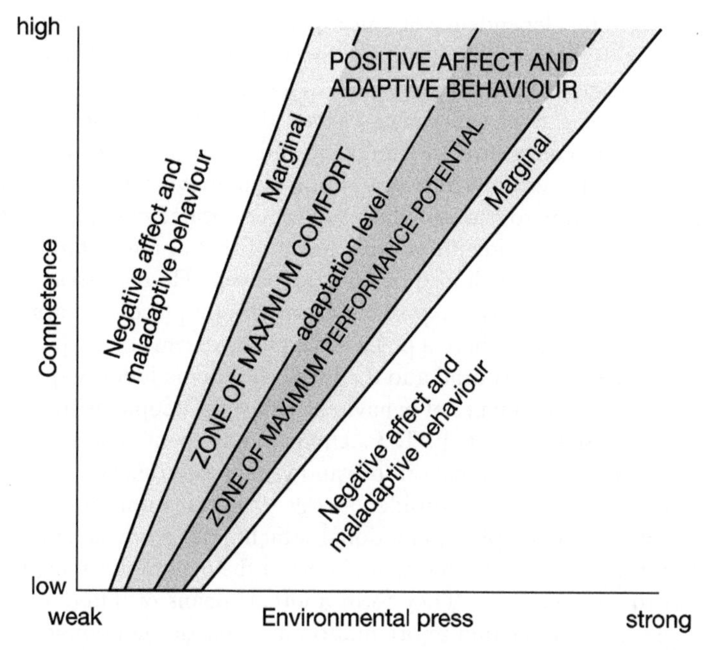

Figure 14.1 Ecological model (source: Lawton. M. P. and Nahemow, L. (1973) 'Ecology and the aging process', in Eisdorfer, C. and Lawton. M. P. (eds) *Psychology of Adult Development and Aging*, Washington, DC: American Psychological Association)

tional approach concerned with accessibility, mobility and sensory awareness within space and how this relates to building development (Hanson, 2001; Kelly, 2001; Peace and Holland, 2001a). Understanding how older people face environmental press with the aid of assistive technology is currently underpinning multi-disciplinary research, particularly between architectural, engineering, computing and social disciplines (Fisk, 2001; Marshall, 2001). Yet both subjective (meaning) and objective (function) factors will influence how people interact with, feel about, or create affective bonds with different aspects of environment. Our understanding of the impact of environment on identity in later life relates to interactions between the material, the social and the psychological.

Spatiality in later life – individual and structural

Another important area of influence has been identified by critical social geographers working in gerontology at the levels of both individual experience and societal structure. Here we cite two authors – Graham Rowles and Glenda Laws. First, the seminal work of Graham Rowles offers an early geographical perspective. In *Prisoners of Space?* (1978) Rowles worked towards a theoretical perspective on the geographical experience of older people, in terms of both space and place. He uses in-depth biographical material from five older people to show the inter-dependency of four aspects of experience relating to activity, orientation, feeling and fantasy, which enables a holistic framework to be created for each individual's environmental transactions.

The four aspects of experience display their own complexity. *Activity* has three attributes constituted by frequency and its more or less habitual nature. For example, this would range from immediate movement around the proximal physical setting to occasional trips to visit relocated children. In relation to *orientation* Rowles also identifies three levels of spatial understanding (see Box 14.1[1]); the third of which also has five sub-levels. He discusses awareness of place as hierarchical between proximate and distant places, but argues that the order can be changed where a place has a particular meaning (p. 171). Orientation may also be territorial and the levels of schema identify spaces that are prescriptive of certain types of behaviour with some people being more possessive of space than others (p. 173). Understandings of spaces could also be expanded through additional sensory and virtual experiences.

Then he moves on to consider *feelings* – the meaning attached to space which helps to define place, emotional attachments from 'dread' to 'elation' (see Box 14.2). These are an essential part of the geographical experience but difficult to communicate. They form another means of differentiating space and there is an association to orientation in terms of the boundaries between schemata coinciding with changing or contradictory emotional attachments.

Finally, Rowles turns to 'vicarious forms of geographical experience' – *geo-*

Box 14.1 Orientation

Orientation relates to an understanding or cognition of space

Personal schema – facility to direct action within 'lived space'
Specific schemata – paths, routes, nodes used to negotiate the environment and facilitate orientation in different circumstances, e.g. seasonal, temporal, gradational
General schema – related to experientially distinctive domains embedded within each other:

- *Home* – the fulcrum
- *Surveillance zone* – field of vision from the home, not inviolable space but watchful space
- *Neighbourhood* – functionally defined but with experiential boundaries, social affinity, identification threatened by neighbourhood transition
- *City* – more fragmented and disjointed differentiation of space; separate from local; known through personal history; individual settings within wider spatial entity; affection plus transitions
- *'Beyond spaces'* – significant locations

Box 14.2 Feelings

Feelings classified in terms of *degree of permanence* as:

Immediate – situation-specific; reactive emotions
Temporary – but potentially repetitive, e.g. diurnal, seasonal, contradictory
Permanent – intimate emotions bringing stability, attachment; subjective assessments

Feelings classified in terms of *degree of association* as:

Personal – individual, unique experiences of physical, social context
Shared – mediation of others in experience; involvement in common context; consensus of feelings; commonality of feelings

graphical fantasy through the places of imagination – experienced in recollection and imagination (see Box 14.3). He says that 'fantasies are creative "grand fictions" through which the individual moulds a separate personal reality' (p. 180) and stresses the great significance that he feels is played by participation in both past and displaced contemporary settings. He notes, 'reflective fantasy imbued contemporary place with depth of meaning inasmuch as the stream of its past was incorporated within its experienced identity' (p. 183) and he talks of this experience as being liberating, allowing for an inclusivity that is engaging. This modality is seen to encompass the other three and Rowles demonstrates through his research how contextual cues such as artefacts, contexts such as rooms or streets, media and, to a lesser degree, people can stimulate geographical fantasy.

By exploring these four aspects of geographical experience Rowles suggests that quality of life is supported where personal capabilities and environmental opportunity are balanced through ongoing adjustments (Rowles, 1978: 190), a view that relates to the work of Lawton and colleagues.

Box 14.3 Fantasy

Fantasy within displaced milieu classified as:

Contemporary milieu – present time
Not contemporary milieu – past, future time
Reflective – reconstruction of biography in consciousness; place becomes wider/larger; provides support for personal identity
Projective – vicarious participation in settings spatially and temporally beyond contemporary environment, e.g. the world of their children; the future world

In concluding his earlier work Rowles discussed the changing emphasis that older people register within and among the modalities of geographical experience. He says,

> These changes involve constriction, selective intensification, and expansion. Constriction refers to changes tending towards geographical lifespace limitation and closure: it is primarily manifest in the realm of action. Selective intensification, involving increasing psychological investment within particular places, is postulated as the dominant motif of change within the realms of orientation and feeling. Finally, expansion refers to changes in which the geographical experience of the older person is enriched. It is suggested that, with advancing age, there is an expansion of the role played by fantasy in the totality of geographical experience.
>
> (Rowles, 1978: 196)

Over time Rowles developed his understanding of geographical experience within rural environments. He developed the concept of 'insideness' (Rowles, 1983b), which is seen to have three attributes:

- *physical insideness* – familiarity with the physical setting,
- *social insideness* – integration with social fabric of community and possible age peer group culture and
- *autobiographical insideness* – based on time and space, a historical legacy of life lived within a particular environment and of importance to developing 'communities of concern' for older people living in rural areas.

More recently, Rowles has discussed the dynamic transformation of spaces into places to forge the personal sense of 'being in place' (Rowles, 1991; 2000) and Rowles and Watkins (2003) have identified elements of the art of place-making as 'history, habit, heart and hearth', moving towards a life-course model of environmental experience.

The second author considered here is Glenda Laws,[2] who developed issues concerning the spatiality of ageing from a structural perspective. She saw spaces and places as age-graded, recognising the ageism of space where 'youth is everywhere' (Laws, 1997) and utilised the work of cultural geographers and sociologists concerning the spatiality of social life within gerontology (Zukin, 1992; Featherstone and Lash, 1995; Pile and Thrift, 1995) to indicate how space can be used to structure society across time.

There are links between social processes and spatial arrangements in society that can be seen to operate at both the individual and collective levels. A particular example used by Laws is the development of 'specialised' accommodation for older people from the evolution of the workhouse into the home for the aged in the UK, to the growth of retirement communities in the USA. She shows how the history of accommodation for the 'poor, unemployed and old' led to segregated living in workhouses – a form of age-excluded spatialisation (for British development see Willcocks *et al.*, 1987; Peace *et al.*, 1997). She argues that the creation of nursing and care homes changed attitudes about the locus of care for certain groups of older people – 'the webs of social relations in which we are enmeshed are constituted through space' (Laws, 1992: 92) – and supported this argument by considering how the family home, once the centre for family relations, is held in memories – as fantasy space in Rowles' terms.

Laws sees identity as 'fluid and constantly renegotiated'. She says that identities may be imposed on the subject by external forces (stereotypes – ageist, racist, sexist) or they can be self-nominated or internal forces – acceptance, internalisation of stereotypes. She argues that there are spatial dimensions of identity formation and that our 'identities are structured as much by material acts as they are by discursive practices' (Laws, 1997: 93).

Discourses are created within material contexts. 'What roles do space and place play in the fabrication, reconstruction, celebration or repudiation of social identities? . . . Where we are says a lot about who we are' (p. 93). In other words, that who we are in the domestic setting, through exercise of day-to-day influence and control – through objects, routines, clothes, choice of media – is consequent upon capital resources and thus to society beyond the home.

Laws argued that there are several dimensions of spatiality which affect identity formation and which influence the inclusion or exclusion of older people within society. In particular she considers

- *accessibility* – citizenship status from room access to nationhood,
- *mobility* – a marker of your position relative to others,
- *motility* – the body's 'potential' to move – labelling of frailty,
- *spatial scale* – how older people are seen as individuals within intimacy of family or as a member of a social group – the 'burden' of older people and
- *spatial segregation* – power issues, seen as a form of social control of groups.

She analyses the domestic built environment to consider the spatialities of age relations through the separation of generations between households, and distance and change over historical periods, and considers that the experience of being old varies according to your environment. Space can mediate or constrain personal experience. Poorhouses evolved into the 'homes for the aged', and retirement communities are sold as places for active, happy people, 'busy while relaxing'. Significantly, location is important in spatially segregating older people from big-city life, which nonetheless remains within reach. She calls this a form of 'discursive identity'.

Attachment to place

Examination of both macro and micro levels of environmental interaction/transaction finally focuses on attachment to place, of importance in maintaining self through the continuity of history, and in providing a shield of support in times of change (Altman and Low, 1992). The significance of place at different points in the life course has been discussed by many authors (e.g. Howell, 1983; Marcus, 1995; Zingmark *et al.*, 1995) Whereas Bowlby (1953) saw attachment as behaviour related to early life experiences, Rubinstein and Parmelee (1992) viewed attachment behaviour as a life-course phenomenon concerning place as lived currently and in memory. They also differentiated between spaces that can be adapted or manipulated to support competence and how they can evolve to the more complex meaning of place: 'Whereas a space is any defined piece of territory, a place has personal significance, a significance established through time spent in or with the space' (p. 142) and '[p]ersonal experience, either direct or vicarious . . . and social interaction lead the person to attach meaning to a defined space; as a result, within his or her own identity, it becomes a place' (p. 142).

Their model of place attachment in later life shows how the definition of place relates to both time and space: 'the very notion of place implies a conflation of space and time such that attachment to a particular place may also represent attachment to a particular time' (p. 142). This is sensed in the energy and emotion that people bring to the description of spaces which may range from an individual object as a reflection of place through to a country with particular meaning for the individual. The language of the space will define a 'place' and Rubinstein argues that this attachment may be 'strong or weak', 'positive or negative', 'narrow, wide or diffuse' (Rubinstein, 1990).

Rubinstein and Parmelee present a model of place attachment in late life and argue that there are three essential elements for understanding: geographic behaviour, identity and inter-dependence. These encompass the locational and social life of the individual and provide for the 'dialectic of autonomy and security' (Parmelee and Lawton, 1990). The three constructs can be viewed as spanning two dimensions: *the collective* (i.e. the meanings inherent to a given culture and shared by its members) and *the individual* (i.e. meanings derived from personal attitudes, beliefs and experiences). They argue that place attachment is important to older people for three reasons:

- keeping the past alive,
- remaining constant during times of change
- maintaining a sense of continued competence,

and stress the importance of bringing together the 'significance of objective place characteristics and subjective place experiences' (Rubinstein and Parmelee, 1992). In other words, whether the place allows you to do what you want to do or whether an alternative place would be better, and who decides. As Rubinstein and Parmelee acknowledge, this set of arguments has parallels with those of Rowles, outlined above.

In a more complex examination of attachment, Rubinstein (1989) has also been influential in examining the micro-environment of the home. In taking a psycho-social approach, he attempted to theorise the connections between the home environment (defined as both the dwelling place and the objects within it), and the psychological integration of home. From an intensive study of a small group of older Americans, he described three classes of empirically derived psycho-social processes. These were defined as relating to socio-cultural order ('social-centered processes'), the life course ('person-centered processes') and the body ('body-centered processes'). In Rubinstein's view, the recreation of shared (cultural) notions of correct standards and behaviour, and the proper times and places for particular activities and so on, is a fundamental part of the processes that link person to place. He sees socio-cultural ordering as the 'process through which the individual interprets and then puts into practice for herself the version of a collective sense of what is proper and what it means to be a person' (Rubinstein, 1989: S48).

Within person-centred processes, Rubinstein considered four elements of place attachment involving increasing degrees of identification with objects in the environment – from simple accounting for the province of environmental features to a close identification with aspects of the environment, the ultimate environmental press:

> embodiment can be important to some older people for whom, at the current time, aspects of the physical environment may have a potential for greater endurance than their own bodies. Environmental features may therefore be assigned the task, through embodiment, of carrying the load of personal meaning and thereby aid in the maintenance of self, when it is threatened.
>
> (Rubinstein, 1989: S50)

In terms of the body, he described two processes: entexturing, where regulation of the environment serves to induce a sensory state of comfort; and environmental centralisation, so that the environment is manipulated over time to accommodate increasing limitations of the body. Peripheral areas are closed off and living space is concentrated into central zones; less important activities are abandoned to muster energy for those that are felt to be important.

Each of these authors has influenced our work, which builds on, expands and challenges some of their ideas, their theoretical and conceptual developments. But first we consider our methodological approach.

Methodological approach

Over the past 25 years we have, at different times, been part of a multidisciplinary team of researchers[3] who have studied living environments for older people both 'ordinary' (everyday housing) and 'special' (collective accommodation with care services). In many of these studies, 'quality of life' has been an underlying concept and we have sought to relate personal issues of satisfaction, psychological well-being and morale to the domain of environment in order to understand psycho-social interactions which may lead to either positive or negative experience.

Research to date has involved a range of quantitative and qualitative research methods that have to some extent been influenced by the remit of funders (see Willcocks *et al.*, 1987; Holland, 2000; Peace and Holland, 2001b; Hanson *et al.*, 2004). This has included the use of psychometric measures and environmental assessments within semi-structured surveys; in-depth qualitative studies of settings involving participant observation; in-depth interviewing developing housing histories; cognitive mapping; collection of photographic records of domestic and special housing interiors, floor plans and object inventories; and the use of video (Peace, 1977; Holland, 2001; Hanson *et al.*, 2004).

During these studies we have remained alert to the ethnographic approaches adopted by Rowles and Rubinstein, who have both based much of their work on in-depth qualitative studies with small samples of respondents with whom they have spent a great deal of time. Such studies indicate the importance of understanding the complexity of person–environment interaction at a detailed level. While 'quality of life' has also been a central focus of the Growing Older programme funded by the Economic and Social Research Council, we have chosen to pursue a more ethnographic and participative route when looking at 'Environment and Identity in Later Life' in a study across settings (Peace *et al.*, forthcoming).

Our research, concerned with the relationship between environment and identity, makes self-perception or the factor 'ego strength' central. A recognition that people experience multiple identities highlights both the societal and personal factors which impinge upon the individual. For example, financial resources, social networks and incidents of age discrimination need to be seen alongside the experiences of an ageing individual of increasing co-morbidity, social isolation and sensory loss. We have been concerned with people's interactions with environments at both objective and subjective levels. As a consequence, our ethnographic study using participative research techniques focused on 'lives of quality' rather than the more usual conceptualisation of 'quality of life' (Peace *et al.*, forthcoming).

The late 1990s saw the beginning of greater participation by research subjects within the research process. This movement has evolved in two ways: first through certain groups – particularly younger people with disabilities – who see the research process as emancipatory and empowering; and second through the move by policy-makers to involve service users in the development of services and policies, sometimes involving research skills. Within the UK, some older people have become involved in the research process (Peace, 2002) and the present research team was concerned to develop approaches led by the views of older people. Consequently the research started with older people themselves and focus groups were held to establish what they thought was most significant about the places where they lived. From this information and previous research findings comprehensive interview schedules were developed, including a new

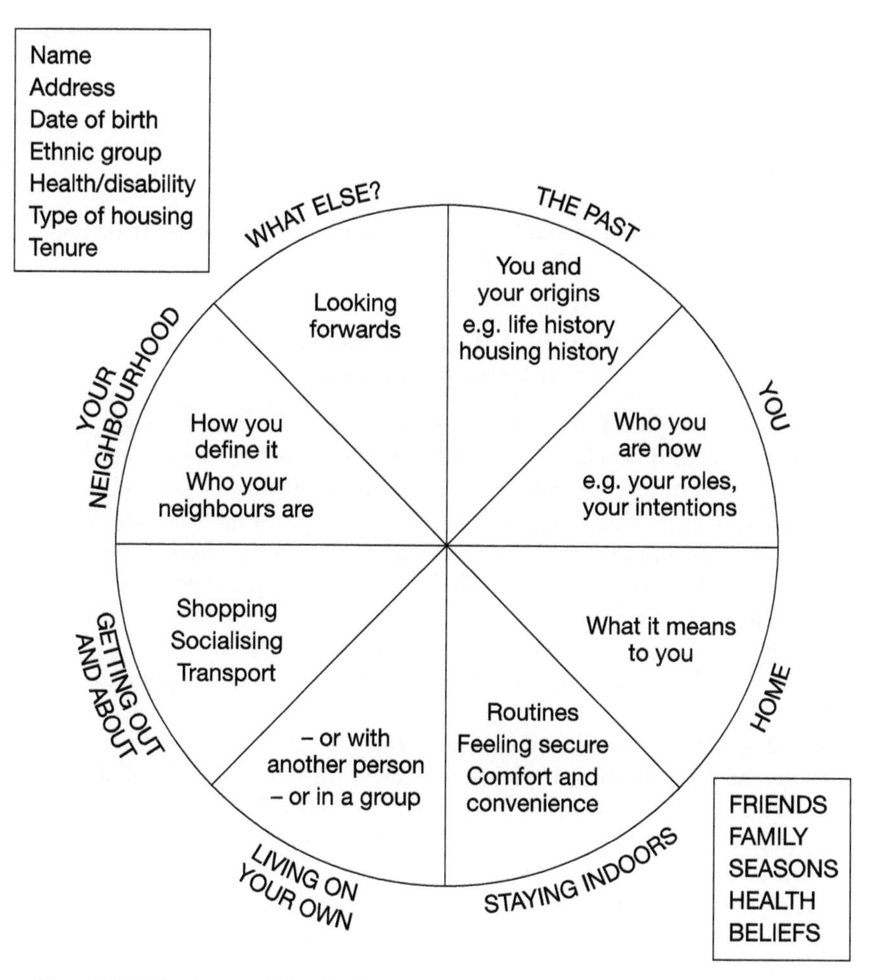

Figure 14.2 The facets of life wheel (source: Peace, S. M., Holland, C. and Kellaher, L. (2003) *Environment and Identity in Later Life: a cross-setting study*)

research tool – a prompt for discussion in the form of a rotating 'facets of life wheel' (see Figure 14.2).

Twenty men and 34 women aged between 61 and 93 years were interviewed in depth from three locations (rural to metropolitan). Five of these were also included in a video about aspects of attachment to place. The respondents lived in their own homes – both domestic and non-domestic, rented and owned. The inclusion of both domestic and non-domestic settings makes this data set unique and the term 'home' is used inclusively. The homes where the 54 respondents lived included

- 6 rooms in residential care homes,
- 6 flats in sheltered housing,

- 8 bungalows,
- 4 terraced houses,
- 10 flats including two high-rise,
- 12 semi-detached houses and
- 8 detached houses including farmhouses.

Content analysis of data from both the focus groups and the individual cases form the basis of findings discussed below.

The relationship of environment to identity: a British example

This research builds on the work of those theorists discussed above, arguing that a life of quality is achieved when the older person can adopt strategies that allow attachment or connection to the social and material fabric of everyday life. Peace *et al.* (forthcoming) see this attachment as a material and psychological process that accumulates across the life course. The research shows how people constantly generate, reinforce and dissolve connections with environment taken in its widest sense. In what we have called a theory of re-engagement for a life of quality, we examine what appears to be an essential comfort level for environmental connectedness or performance potential which people strategically seek to maintain by creating a particular balance between self and society, reflexivity and reflectivity, inside and outside. This engages with the mechanisms that people use to achieve the 'adaptation level' described by Lawton and Nahemow (1973; see also Lawton, 1999; Nahemow, 2000) and incorporated into the ecological theory of ageing (see Rubinstein and De Medeiros, 2004), and with Rowles' concern with ongoing adjustment to place.

The older people interviewed employed sets of strategies to optimise the mix of 'self' and 'other', using material, social and cognitive mediators – including household objects and the dwelling itself, associative life and place memories. Some people find that over time the number and quality of their attachments to environment become compromised by decline in their own competence and/or the demand characteristics of environment. The point at which these attachments become insufficient for tolerable living equates to the point where adaptive behaviour cannot rebalance environmental press. To avoid or circumvent this, older people look for ways to reinforce and increase their points of attachment to place, finding and sometimes creating new options to enable them to do so. We have called this process 'option recognition'. The interaction between person and environment can include modification or adaptation of behaviour or environment, using formal and informal services or relocation; and cognitive or imaginative processes replacing physical action (see Peace *et al.*, forthcoming). It may lead some people to live in a new space (accommodation) within a valued place (location).

A life of quality is one where mastery is maintained over the strategies necessary to make attachments or connections – to have the capacity to attach to social and material 'fabric' – at a level, above basic self-maintenance, where the individual can feel secure about her/his own identity. With their chosen options, older people may engage or disengage with particular activities, things, people or interests. Interaction may be at the micro level (e.g. with collections of objects) and the macro level (overseas travel). Positive engagement can range from third-age activism to willing disengagement (Tornstam, 1989). But engagement levels can also be reversed, for example after a change of environment or health status, as older people willingly or unwillingly reassess their ability to maintain the environmental status quo. Timing may be critical to the relative success of strategies for relocation and re-engagement, for example in the case of relocation to residential care settings (temporary or permanent). Here age integration may have more or less importance through levels of control.

Identity and the home – space, objects, attachment

In the private space of their homes, our respondents had created the richly encoded and materially complex worlds that have also been recorded elsewhere (Sixsmith, 1986; Hanson *et al.*, 2004). Respondents living in residential homes necessarily had simplified personal spaces but these could be enhanced to some extent by the material and human complexities of the communal day spaces in their homes. However, a life of quality has to be finely balanced against aspects of institutionalisation, and interaction within and between public and private spaces is just as crucial to self-identity within care settings (Willcocks *et al.*, 1987) as in domestic settings.

Most of our respondents were living in non-age-related housing in mixed communities, either alone or as couples. This generally meant that they had much more home space (including outside storage and gardens) than for example people in sheltered housing or residential care homes, but very few of them considered that that they had too much space. Guest bedrooms were very important, especially for grandchildren and children, and other spaces were used for hobbies, storage and personal 'expansion' space (to move around within the house). Most people in all types of accommodation had a favourite or most used space within the home, but it was important for them to be able to move away from that anchor point, often as a part of their daily routine, in order to return to it re-energised by the change. For some people this took them out of the house completely, for others the 'journey' might be just into another room, but in all cases the action of moving between points had physical and psychological benefits.

Respondents explained the significance and provenance of many objects in their homes, especially pictures, photographs and memorabilia. These included representations of important others (kin and non-kin) and respected figures, representations of significant other places (often related to previous homes, childhood, or past occupations), craft work and representational images connected to personal interests or achievements or those of other family members.

They often also included objects that the respondent had no particular attachment to but were displayed to please other people. These objects and the manner of their display were representing *self to self* (to remind, to reinforce selected aspects of personality), *self to others* (claims about personality and achievements) and *others to self* (visual reminders of absent people, including changing images over time, e.g. of grandchildren). While many or most of these items individually were not indispensable to the respondent's well-being, as a set they strongly underpinned the sense of identity and ontological security and in our opinion are a primary focus of attachment related to the life of quality.

Gardens, yards and landscapes: semi-public, semi-private space

The boundaries of space are crucial for some people and our research begins to show the importance of the intervening space between accommodation and street life. This may extend across the range of private to public space from the secluded garden to the grassed surround segregating the residential home. The importance of the natural world from a long-standing garden where plants are linked to people, to the 'pots' that can be tended, begins to reveal a relatively untapped area of data concerning environment and identity, and one that has implications for the ways in which older people perceive and use public open spaces.

Neighbourhood and identity

As people move around the neighbourhood, they negotiate and interpret the various elements it presents. Whether simple or highly complex, material environments can present barriers to social engagement and the sense of control over movement or mobility can be crucial to self-esteem. For some people, ageing entails a heightening of the sensual impact of the micro environment as material elements such as ground textures, changes in level and barriers become more problematic. Transport in and out of the neighbourhood is an issue for many. Some people, especially the oldest women, have never had independent access to a car, and as they age people are increasingly likely to give up driving, so that moving out beyond the neighbourhood requires accessible public transport, taxis or the help of others.

Proximate neighbourhoods (i.e. within walking distance) were very important to most of our respondents (with the exception of those in residential care homes). The tension between comfort and discomfort in neighbourhoods emerged as an important theme (see Peace, Holland and Kellaher, forthcoming). No longer being able to go out independently is a critical stage in identity construction because, without the wider contexts that lie beyond the dwelling, the home itself becomes diminished as a source of identity construction. Continued capacity to engage with 'the other' is represented by neighbourhood in a way that the immediate domicile cannot demonstrate or prove. For a majority of our respondents, actual engagement in material and social neighbourhoods was still essential to well-being and self-identity, confirming the spatiality of identity described in Laws' work. Our research suggests it is essential that policies on regeneration and development recognise how older people locate themselves within community life.

Time, place and space

Our research covered both temporal and spatial aspects of respondents' housing histories. Within each location we have a group of people with lifelong, if intermittent, connections to that location (broadly defined) for whom their present home represents a spatial continuity or 'homecoming' that is location-specific, and another group for whom the specific location is secondary to the particularity of the actual dwelling place and/or the social milieu. These differences have an effect on the extent to which people identify with the places where they live and on their decision-making processes as they consider the possibility of future moves.

In most cases the respondents' present tenure and quality of housing emerged from their personal and housing histories with trajectories to rented or owner-occupied housing determined at an earlier phase of life and wealth accumulation related to the timing and location of their moves through their housing careers. As in other studies, the exceptions tended to be people who had suffered major disruptions through divorce, financial catastrophe or health crisis. The history behind their present accommodation was one aspect of the personal meaning of their present housing status. As a result of socio-economic and ideological influences, the extent to which the personal identity of the respondents was vested in the status of their homes was both complex and likely to be different to patterns among the next several cohorts of British older people.

Making space for identity

To conclude, present analysis of the recently completed British study of 'Environmental and Identity in Later Life' confirms and extends the seminal work outlined at the start of this chapter. Place identity is a component of self-identity for older people living in Western developed countries that may be experienced differentially in terms of micro and macro environments.

Our data indicate that understanding the meaning of 'home' as context is crucial to person–environment interaction in later life at the micro level. Individual role identity will affect attachment to 'home' as a more or less valued place within the life course, and this will be enhanced by social interaction which may, or may not, be familial. We do not claim that 'home' is always positive, and acknowledge the importance of control over private space as a crucial indicator of selfhood. Negative associations can stem from denial of self-actualisation within settings domestic and non-domestic. Our study reinforces the centrality of mastery over the micro environment to develop a place with a degree of comfort that allows for well-being. Developing enabling environments through adaptations – encompassing the home-made to sophisticated assistive technology – is to be welcomed because it can increase personal control.

However, place identity is also seen as historic and wide-ranging. Respondents who had migrated as adults to Britain commented on their own place identity 'in limbo' between cultures. Our research indicates the importance of housing history within place attachment and we confirm Rubinstein's emphasis on 'keeping the past alive' and the temporal context within Rowles' description of levels of spatial experience.

Such discussions are necessary for a greater understanding of current macro environments – from street-level to country-level – where older people may experience different levels of inclusion and exclusion. On the one hand there is a certain homogeneity about ageing in British and other 'Western' societies that collude in a spatial ageism, so that, for example, most older people are reluctant to go out at night. A minority of older people have dealt with issues of inclusivity by moving into age-related communities but, for most, strategies have to be found in the wider community. On the other hand, the increasing diversity of older people adds complexity to power relations and social interactivity and the exercise of option recognition. A majority of older people continue to search for ways to be involved in society, albeit often at a less stressful pace than previously. Our theory of re-engagement stresses this involvement and we argue that in the processes of attachment described over decades by the authors discussed here, both the activities and results of maintaining involvement are crucial to well-being and the life of quality in old age.

Notes

1 Boxes 14.1–3 are developed by Sheila Peace from Rowles, G. (1978) *Prisoners of Space?*, pp. 157–203.
2 Sadly, Glenda Laws died in 1996.
3 Research based at both the Centre for Ageing and Environmental Studies (CESSA) – founded at the Polytechnic of North London (now London Metropolitan University) in 1983 and through the Centre for Ageing and Biographical Studies (CABS), School of Health and Social Welfare, The Open University.

REFERENCES

Abel, Emily K. (1992) Parental dependence and filial responsibility in the nineteenth century: Hial Hawley and Emily Hawley Gillespie, 1884–1885, *The Gerontologist*, 32: 519–26.

Abrams, Philip and Bulmer, Martin (1985) Policies to promote informal social care: some reflections on voluntary action, neighbourhood involvement, and neighbourhood care, *Ageing and Society*, 5: 1–18.

Achenbaum, W. A. (1995a) *Crossing Frontiers: Gerontology emerges as a science*, Cambridge: Cambridge University Press.

Achenbaum, W. A. (1995b) Images of old age in America, 1790–1970: a vision and a re-vision, in M. Featherstone and A. Wernick (eds) *Images of Aging: Cultural representations of later life*, London: Routledge.

Achenbaum, W. A. and Kusnerz, P. A. (1978) *Images of Old Age in America, 1790 to the present*, Ann Arbor: Institute of Gerontology.

Adamek, Margaret E. (1992) Should the government pay? Caregiver views of government responsibility and feelings of stigma about financial support, *Journal of Applied Gerontology*, 11, 3: 283–97.

Age Concern (2003) http://www.ageconcern.org.uk/. Age Concern England. Website accessed 15 October 2003.

American Geriatric Society Panel on Falls in Older Persons (2001) Guidelines for the prevention of falls in older persons, *Journal of the American Geriatric Society*, 49: 664–72.

Anderson, B. (1993) *Imagined Communities: Reflections on the origin and spread of nationalism*, London: Verso.

Anderson, Joan M. (2001) The politics of home care: where is 'home'? *Canadian Journal of Nursing Research*, 33, 2: 5–10.

Anderson, R. N. (2002) *Deaths: Leading causes for 2000*, National Vital Statistics Reports, 50(16), Hyattsville, Maryland: National Center for Health Statistics.

Andrews, G. J. (1997) Private residential care for elderly people: a socio-spatial study of the impacts of care in the community policy in the 1990s, unpublished Ph.D. dissertation, University of Nottingham.

Andrews, G. J. (2002a) Private complementary medicine and older people: service use and user empowerment, *Ageing and Society*, 22: 343–68.

Andrews, G. J. (2002b) Towards a more place-sensitive nursing research: an invitation to medical and health geography, *Nursing Inquiry*, 9, 4: 221–38.

Andrews, G. J. (2003a) Placing the consumption of private complementary medicine: everyday geographies of older people's use, *Health and Place*, 9: 337–49.

Andrews, G. J. (2003b) (Re)thinking the dynamic between health and place: extending theory and focus in therapeutic geographies, *Area*, in press.

Andrews, G. J. and Kearns, R. A. (2004) Everyday health histories and the making of place: the case of an English coastal town, *Social Science and Medicine* (in press).

Andrews, G. J. and Kendall, S. (2000) Dreams that lie in tatters: the changing fortunes of the nurses who left the NHS to own and run residential homes for elderly people, *Journal of Advanced Nursing*, 31, 4: 900–8.

Andrews, G. J. and Phillips, D. R. (2000) Moral dilemmas and the management of private residential care homes: the impact of care in the community reforms in the UK, *Ageing and Society*, 20: 599–622.

Andrews, G. J. and Phillips, D. R. (2002) Changing local geographies of private residential care for older people 1983–1999: lessons for social policy in England and Wales, *Social Science and Medicine*, 55: 63–78.

Andrews, G. J., Gavin, N., Begley, S. and Brodie, D. (2003) Assisting friendships, combating loneliness: Users' views on a befriending scheme, *Ageing and Society*, 23, 3: 349–62.

Angus, Jan (1994) Women's paid/unpaid work and health: exploring the social context of everyday life, *Canadian Journal of Nursing Research*, 26, 4: 23–42.

Antonelli, E., Rubini, V. and Fassone, C. (2000) The self-concept in institutionalized and non-institutionalized elderly people, *Journal of Environmental Psychology*, 20: 151–64.

Arber, Sara (1997) Comparing inequalities in women's and men's health: Britain in the 1990s, *Social Science and Medicine*, 44, 6: 773–87.

Arber, Sara and Gilbert, G. Nigel (1989) Men: the forgotten carers, *Sociology*, 23, 1: 111–18.

Arber, Sara and Ginn, Jay (1990) The meaning of informal care: gender and the contribution of elderly people, *Ageing and Society* 10: 429–54.

Arber, Sara and Ginn, Jay (1995) Connecting gender and ageing: a new beginning? in S. Arber and J. Ginn (eds) *Connecting Gender and Ageing: A sociological approach*, pp. 73–8, Buckingham: Open University Press.

Arber, Sara and Ginn, Jay (1993) Gender and inequalities in health in later life, *Social Science and Medicine*, 36, 1: 33–46.

Argent, N. M. and Rolley, F. (2000) Financial exclusion in rural and remote New South Wales: a geography of back branch rationalisation, 1981–1998, *Australian Geographical Studies*, 38: 182–203.

Armstrong, Hugh and Armstrong, Pat (2002) *Thinking it Through: Women, work, and caring in the new millenium.* Halifax: Nova Scotia: Advisory Council on the Status of Women, Maritime Centre for Excellence in Women's Health and a Healthy Balance.

Armstrong, Pat and Armstrong, Hugh (1999) *Women, Privatization, and Health Care Reform: The Ontario case*, Ottawa: Health Reform Reference Group, Centres of Excellence for Women's Health Program, Women's Health Bureau, Health Canada.

Armstrong, Pat and Kits, Olga (2001) *One Hundred Years of Caregiving*, Ottawa: Law Commission of Canada.

Aronson, Jane (1990) Women's perspectives on informal care of the elderly: public ideology and personal experience of giving and receiving care, *Ageing and Society*, 10: 61–84.

Ashley, A. and Bartlett, H. (2001) An evaluation of a walking scheme based in primary care: the participants' perspective, Primary Health Care Research and Development, 2: 98–106.

Ashton, J. and Seymour, H. (1988) *New Public Health: The Liverpool experience*, Milton Keynes: Open University Press.

REFERENCES

Astley, T. (1995) *Coda*, London: Secker and Warburg.

Australian Institute of Health and Welfare (2001) *Chronic Diseases and Associated Risk Factors in Australia*, (Canberra: AIHW (Cat. no. PHE–33).

Australian Institute of Health and Welfare (2002) *Older Australia at a Glance*, Canberra: AIHW.

Averill, J. B. (2002) Voices from the Gila: health care issues for rural elderly in South-Western New Mexico, *Journal of Advanced Nursing*, 40: 654–62.

Bachelard, G. (1994) *The Poetics of Space*, trans. M. Jolas, Boston, MA: Beacon Press.

Bagilhole, B. (1996) Tea and sympathy or teetering on social work? An investigation of the blurring of the boundaries between voluntary and professional care, *Social Policy and Administration*, 30: 189–205.

Baillon, S., Scothern, G., Neville, P. and Boyle, A. (1996) Factors that contribute to stress in care staff in residential homes for the elderly, *International Journal of Geriatric Psychiatry*, 11, 3: 219–26.

Baltes, P. B. and Baltes, M. M. (1990) Psychological perspectives on successful aging: the model of selective optimization with compensation, in P. B. Baltes and M. M. Baltes (eds) *Successful Aging: Perspectives from the behavioral sciences*, pp. 1–34, New York: Cambridge University Press.

Baltes, M. M. and Carstensen, L. L. (1999) Social-psychological theories and their applications to aging: from individual to collective, in V. L. Bengston and K. W. Schaie (eds) *Handbook of the Theories of Aging*, New York: Springer.

Baltes, M. M. and Carstensen, L. L. (1996) The process of successful ageing, *Ageing and Society*, 16: 397–422.

Baltes, M. M., Wahl, H. W. and Schmidt-Furross, U. (1990) The daily life of elderly German activity patterns, personal control, and functional health, *Journal of Gerontology: Psychological Science*, 45: 173–9.

Baltes, P. B. and Baltes, M. M. (1990) Psychological perspectives on successful aging: the model of selective optimisation with compensation, in P. B. Baltes and M. M. Baltes (eds) *Successful Aging: Perspectives from the behavioural sciences*, Cambridge: Cambridge University Press.

Baltes, P. B. and Mayer, K. U. (eds) (1999) *The Berlin Aging Study: Aging from 70 to 100*, New York: Cambridge University Press.

Baltes, P. B. and Smith, J. (1997) A systematic-holistic view of psychological functioning in very old age: introduction to a collection of articles from the Berlin Aging Study, *Psychology and Aging*, 12: 395–409.

Baltes, P. B. and Smith, J. (2002) New frontiers in the future of aging: from successful aging of the young old to the dilemmas of the fourth age. Plenary Lecture prepared for the Valencia Forum, Spain, April 1–4.

Bamford, C., Gregson, B., Farrow, G., Buck, D., Dowswell, T., McNamee, P., Bond, J. and Wright, K. (1998) Mental and physical frailty in older people: the costs and benefits of informal care, *Ageing and Society*, 18, 3: 317–54.

Bandura, A. (1989) Social cognitive theory, *Annals of Child Development*, 6: 3–58.

Barbotte, E., Guellemin, F., Chau, N. and the Lorhandicap Group (2001) Prevalence of impairments, disabilities, handicaps and quality of life in the general population: a review of recent literature, *Bulletin of the World Health Organization*, 79, 11: 1047–55.

Barnes, S (2002) The design of caring environments and the quality of life of older people, *Ageing and Society*, 22: 775–89.

Bartlett, H. (1999) Primary health care for older people: progress towards an integrated strategy, *Health and Social Care in the Community*, 7, 5: 342–9.

Bartlett, H. and Phillips, D. R. (1995) Ageing in the United Kingdom: a review of demographic trends, recent policy developments and care provision, *Korea Journal of Population and Development*, 24, 2: 181–95.
Bartlett, H. and Phillips, D. R. (1996) Policy issues in the private health sector: examples from long-term care in the UK, *Social Science and Medicine*, 43, 5: 731–7.
Bartlett, H. and Phillips, D. R. (1997a) Ageing and aged care in the People's Republic of China: national and local issues and perspectives, *Health and Place*, 3: 149–59.
Bartlett, H. and Phillips, D. R. (1997b) Age, in M. Pacione (ed.) *Britain's Cities*, London: Routledge.
Bartlett, H. and Phillips, D. R. (2000) The United Kingdom: demographic trends, recent policy developments, and care provision, in V. L. Bengston, K.-D. Kim, G. C. Myers and K.-S. Eun (eds) *Aging in East and West: Families, states and the elderly*, pp. 169–89, New York: Springer.
Bass, David, Noelker, Linda and Rechlin, Linda (1996) The moderating influence of service use on negative caregiving consequences, *Journal of Gerontology*, 51, 3: S121–S131.
Basting, A. (1998) *The Stages of Age: Performing age in contemporary American culture*, Ann Arbor: University of Michigan Press.
Basting, A. (2000) Performance studies and age, in T. Cole, R. Kastenbaum and R. Ray (eds) *Handbook of the Humanities and Aging*, 2nd edn, pp. 258–71, New York: Springer Publishing.
Bauer, Keith A. (2001) Home-based telemedicine: a survey of ethical issues, *Cambridge Quarterly of Healthcare Ethics* 10: 137–46.
Bauer, M. (1999) The use of humour in addressing the sexuality of elderly nursing home residents *Sexuality and Disability*, 17, 2: 147–55.
Baum, F. (1993) Noarlunga Healthy Cities Pilot Project: the contribution of research and evaluation, in J. K. Davies and M. P. Kelly (eds) *Healthy Cities: Research and practice*, London: Routledge.
Bauman, Z. (1992) *Mortality, Immortality and Other Life Strategies*, Cambridge: Polity Press.
Bauman, Z. (1996) From pilgrim to tourist – or a short history of identity, in S. Hall and P. du Gay (eds) *Questions of Cultural Identity*, pp. 18–36, London: Sage.
Beck, U. (1992) *Risk Society: Towards a new modernity*, London: Sage.
Bedford, R. and Heenan, L. D. B. (1987) The people of New Zealand: reflections on a revolution, in P. G. Holland and W. D. Johnston (eds) *Southern Approaches: Geography in New Zealand*, pp. 133–77, Christchurch: New Zealand Geographical Society, Christchurch.
Bedford, R. D., Joseph, A. E. and Lidgard, J. M. (1999) *Rural Central North Island: Studies of agriculture–community linkages*, Hamilton, New Zealand: Department of Geography, University of Waikato.
Bell, P. A., Greene, T. C., Fisher, J. D. and Baum, A. (1996) *Environmental Psychology*, 4th edn, Fort Worth, TX: Harcourt Brace College.
Bellah, R., Madsen, R., Sullivan, W. M., Swidler, A. and Tipton, S. M. (1988) *Habits of the Heart: Middle America observed*, London: Hutchinson Education.
Beresford, P. and Croft, S. (1995) *Citizen Involvement: A practical guide to change*, London: Macmillan.
Berger, P. and Luckmann, T. (1991) *The Social Construction of Reality: A treatise in the sociology of knowledge*, Harmondsworth: Penguin.
Bernard, M. (2000) *Promoting Health in Old Age*, Buckingham: Open University Press.
BiC (Boligtrevsel i Centrum) (1994) Co-housing for senior citizens in Europe, report from EU Conference 'Growing Grey in a Happier Way', Copenhagen: BiC.

Biggs, S., Bernard, M., Kingston, P. and Nettleton, H. (2000) Lifestyles of belief: narrative and culture in a retirement community, *Ageing and Society*, 20: 640–72.

Billig, M. (1987) *Arguing and Thinking*, Cambridge: Cambridge University Press.

Binstock, Robert H. (2000) On the unbearable lightness of theory in gerontology, *The Gerontologist*, 40, 3: 367–73.

Birren, J. E. and Schaie, K. W. (1996) *Handbook of the Psychology of Aging* (vol. 4), San Diego, CA: Academic Press.

Blaikie, A. (1995) Photographic images of age and generation, *Education and Ageing*, 10, 1: 5–15.

Blaikie, A. (1997) Beside the sea: visual imagery, ageing and heritage, *Ageing and Society*, 17: 629–48.

Blaikie, A. (1999) *Ageing and Popular Culture*, Cambridge: Cambridge University Press.

Blanchard-Fields, F. and Chen, Y. (1996) Adaptive cognition and aging (Aging Well in Contemporary Society, Part 2: Choices and Processes), *American Behavioral Scientist*, 39, 3: 231–49.

Blanchard-Fields, F., Jahnke, H. and Camp, C. (1995) Age differences in problem-solving style: the role of emotional salience, *Psychology and Aging*, 10: 173–80.

Bland, R. (1999) Independence, privacy and risk: two contrasting approaches to residential care for older people, *Ageing and Society*, 19: 539–60.

Bochel, M. (1988) Public policy and residential provision for the elderly, *Geoforum*, 19: 667–477.

Bond, J., Farrow, G., Gregson, B. A., Bamford, C., Buck, D., McNamee, P. and Wright, K. (1999) Informal caregiving for frail older people at home and in long-term care institutions: who are the key supporters? *Health and Social Care in the Community*, 7, 6: 434–44.

Bondevik, M. and Skogstad, A. (1998) The oldest old: ADL, social network, and loneliness, *Western Journal of Nursing Research*, 20, 3: 325–33.

Bordo, Susan, Klein, Binnie and Silverman, Marilyn K. (1998) Missing kitchens, in H. J. Nast and S. Pile (eds) *Places through the Body*, pp. 72–92, New York: Routledge.

Bosworth, H. B. and Schaie, K. W. (1997) The relationship of social environment, social networks, and health outcomes in the Seattle Longitudinal Study: two analytical approaches, *The Journals of Gerontology* 52B, 5: 197–205.

Bourdieu, P. (1977) *Outline of a Theory of Practice*, Cambridge: CUP.

Bowlby, J. (1953) *Child Care and the Growth of Love* (2nd edn, 1965), Harmondsworth: Penguin.

Bowlby, Sophie, Gregory, Susan and McKie, Linda (1997) 'Doing home': patriarchy, caring, and space, *Women's Studies International Forum*, 20, 3: 343–50.

Bowling, A. (1993) The concepts of successful and positive ageing, *Family Practice*, 10, 4: 449–53.

Brandstadter, J. and Greve, W. (1994) The aging self: stabilizing and protective processes, *Developmental Review*, 14: 52–80.

Brandtstadter, J. and Renner, G. (1990) Tenacious goal pursuit and flexible goal adjustment: explication and age-related analysis of assimilation and accommodation strategies of coping, *Psychology and Aging*, 5: 58–67.

Brawley, E. C. (2001) Environment for Alzheimer's disease: a quality of life issue, *Aging and Mental Health*, 5, 1: 79–83.

Brea, J. A. (2003) Population dynamics in Latin America, *Population Bulletin*, 58, 1, Washington, DC: Population Reference Bureau.

Brenton, M. (2001) Older people's co-housing communities, in S. Peace and C. Holland (eds) *Inclusive Housing in An Ageing Society*, pp. 169–88, Bristol: Policy Press.
Britton, S., Le Heron, R. and Pawson, E. (eds) (1992) *Changing Places: A geography of restructuring*, Christchurch: New Zealand Geographical Society.
Brody, E. M. and Lang, A. M. (1982) They can do it all: aging daughters with their aging mothers, *Generations*, 7: 18–20.
Brown, L. (1989) Is there sexual freedom for our aging population in long-term care institutions? *Journal of Gerontological Social Work*, 13, 3: 75–93.
Bryant, C. and Joseph, A. E. (2001) Canada's rural population: trends in space and implications in place, *Canadian Geographer*, 45: 132–7.
Bryant, L. L., Corbett, K. K. and Kutner, J. S. (2001) In their own words: a model of healthy aging, *Social Science and Medicine*, 53, 7: 927–41.
Buchignani, N. and Esther Armstrong, C. (1999) Informal care and older Native Canadians, *Ageing and Society*, 19, 1: 3–32.
Bullock, Karen, Crawford, Sybil L. and Tennstedt, Sharon L. (2003) Employment and caregiving: exploration of African American caregivers, *Social Work*, 48, 2: 150–63.
Burnette, D. and Mui, A. C. (1994) Determinants of self-reported depressive symptoms by frail elderly persons living alone, *Journal of Gerontological Social Work*, 22: 3–19.
Butler, R. (1975) *Why Survive? Being old in America*, New York: Harper and Row.
Butler, R. and Parr, H. (1999) *Mind and Body Spaces: Geographies of illness, impairment and disability*, London: Routledge.
Bytheway, B. (1995) *Ageism*, Milton Keynes: Open University Press.
Bytheway, B. (2003) Visual representations of later life, in C. Faircloth (ed.) *Aging Bodies: Meanings and perspectives*, Walnut Creek, CA: Alta Mira Press.
Bytheway, B. and Johnson, J. (1998) The social construction of 'carers', in A. Symonds and A. Kelly (eds) *The Social Construction of Community Care*, London: Macmillan.
Caldwell, J. C. (2001) Population health in transition, *Bulletin of the World Health Organization*, 79, 2: 159–60.
Calkins, M. P. (2001) The physical and social environment of the person with Alzheimer's disease, *Aging and Mental Health*, 5: 74–8.
Campbell, A. J. and Buchner, D. M. (1997) Unstable disability and fluctuations of frailty, *Age and Aging*, 26: 315–18.
Campbell, A. J., Reinken, J. and Allan, B. C. (1981) Falls in old age: a study of frequency and related clinical factors, *Age and Ageing*, 10: 264–70.
Campbell, A. J., Borrie, M. J., Spears, G. F., Jackson, S. L. and Brown, J. S. (1990) Circumstances and consequences of falls experienced by a community population 70 years and over during a prospective study, *Age and Ageing*, 19: 136–41.
Caplan, G., Ward, J., Brennan, N., Coconis, J., Board, N. and Brown, A. (1999) Hospital in the home: a randomized controlled trial, *Medical Journal of Australia*, 170: 156–60.
Carlson, N. R. and Buskist, W. (1997) *Psychology: The science of behavior*, 5th edn, Boston, MA: Allyn and Bacon.
Carolina, M. S. and Gustavo, L. F. (2003) Epidemiological transition: model or illusion? A look at the problem of health in Mexico, *Social Science and Medicine*, 57, 3: 539–50.
Carp, F. M. (1976) User evaluation of housing for the elderly, *The Gerontologist*, 16: 102–11.
Carp, F. M. (1978) Effects of the living environment on activity and use of time, *International Journal of Aging and Human Development*, 9: 75–91.
Carp, F. M. (1987) Environment and aging, in D. Stokols and I. Altman (eds) *Handbook of Environmental Psychology*, pp. 329–60, New York: Wiley.

Carp, F. M. (1994) Assessing the environment, in M. P. Lawton and J. A. Teresi (eds) Focus on assessement techniques, *Annual Review of Gerontology and Geriatrics* (vol. 14), pp. 302–23, New York: Springer.

Carp, F. M. and Carp, A. (1984) A complementary/congruence model of well-being or mental health for the community elderly, in I. Altman, M. P. Lawton and J. F. Wohlwill (eds) *Human Behavior and Environment: Elderly people and the environment* (vol. 7), pp. 279–336, New York: Plenum.

Carstensen, L. L. (1991) Selectivity theory: social activity in life-span context, *Annual Review of Gerontology and Geriatrics*, 11: 195–217.

Carstensen, L. L., Graff, J., Levenson, R. W. and Gottman, J. M. (1996) Affect in intimate relationships: The developmental course of marriage, in C. Magai and S. H. McFadden (eds) *Handbook of Emotion, Adult Development, and Ageing*, pp. 227–47, San Diego, CA: Academic Press.

Carstensen, L. L., Hanson, K. A. and Freund, A. M. (1995) Selection and compensation in adulthood, in R. A. Dixon and L. Backman (eds) *Compensating for Psychological Deficits and Declines: Managing losses and promoting gains.*

Carstensen, L. L., Isaacowitz, D. M. and Charles, S. T. (1999) Taking time seriously: a theory of socioemotional selectivity, *American Psychologist*, 54: 165–81.

Carter, S. E., Campbell, E. M., Sanson-Fisher, R. W., Redman, S. and Gillespie, W. J. (1997) Environmental hazards in the homes of older people, *Age and Ageing*, 26: 195–202.

Cartier, C. (2003) From home to hospital and back again: economic restructuring, end of life, and the gendered problems of place-switching health services, *Social Science and Medicine*, 56: 2289–301.

Casey, B. A. and Yamada, A. (2002) Getting older, getting poorer? A study of the earnings, pensions, assets and living arrangements of older people in nine countries, *Labour Market and Social Policy Occasional Papers No. 60*, Paris: OECD.

Castro, C., Otero-Lopez, J., Freire, M., Losada, C., Saburido, J. L. and Pereiro, D. (1995) A comparative study of strategies for coping with stress in different age groups, *Revista Española de Geriatria y Gerontologia*, 2: 79–84.

Cerrato, I. M. and Fernandez de Troconiz, M. I. (1998) Successful aging. But, why don't the elderly get more depressed? *Psychology in Spain*, 2, 1: 27–42.

Chackiel, J. and Plaut, R. (1996) Demographic trends with emphasis on mortality, in I. A. Timaeus, J. Chackiel and L. Ruzicka (eds) *Adult Mortality in Latin America*, pp. 14–41, Oxford: Clarendon Press.

Challinger, Y., Julious, S., Watson, R. and Philip, I. (1996) Quality of care, quality of life and the relationship between them in long-term care institutions for the elderly, *International Journal of Geriatric Psychiatry*, 11, 10: 883–8.

Challis, D. and Hughes, J. (2003) Residential and nursing home care – issues of balance and quality of care, *International Journal of Geriatric Psychiatry*, 18: 201–4.

Chalmers, A. I. and Joseph, A. E. (1997) Population dynamics and settlement systems: a case study of Waikato, *New Zealand Geographer*, 53: 14–21.

Chaplin, R. and Dobbs-Kepper, D. (2001) Aging in place in assisted living: philosophy vs. policy, *The Gerontologist*, 41: 43–50.

Chappell, Neena (1991) Living arrangements and sources of caregiving, *Journal of Gerontology: Social Sciences*, 46, 1: S1–8.

Chappell, Neena and Blandford, Audrey (1991) Informal and formal care: exploring the complementarity, *Ageing and Society* 11: 299–317.

Charness, N. and Bosman, E. A. (1990) Human factors and design for older adults, in

J. E. Birren and K. W. Schaie, *Handbook of the Psychology of Aging*, 3rd edn, pp. 446–63, San Diego: CA: Academic Press.

Charness, N. and Holley, P. (2001) Human factors and environmental suppport in Alzheimer's disease, *Ageing and Mental Health*, 5: 65–73.

Chatterji, R. (1998) An ethnography of dementia: a case study of an Alzheimer's disease patient in the Netherlands, *Culture, Medicine, and Psychiatry*, 22: 355–82.

Cheers, B. (2001) Globalization and rural communities, *Rural Social Work*, Special Australian/Canadian Issue, December, 28–40.

Cheng, Y. H., Chi, I., Boey, K. W., Ko, L. S. F. and Chou, K. L. (2002) Self-rated economic condition and the health of elderly persons in Hong Kong, *Social Science and Medicine*, 55: 1415–24.

Chipperfield, Judith G. (1994) The support source mix: a comparison of elderly men and women from two decades, *Canadian Journal on Aging/Revue Canadienne du Vieillissement*, 13, 4: 434–53.

Chowdary, U. (1990) Notion of control and self-esteem of institutionalized older men, *Perceptual and Motor Skills*, 70: 731–8.

Christie, S. and Tobias, M. (1998) The burden of infectious disease in New Zealand, *Australian and New Zealand Journal of Public Health*, 22, 2: 257–60.

Clair, J. M. (1990) Old age health problems and long-term care policy issues, in S. M. Stahl (ed.) *The Legacy of Longevity: Health and health care in later life*, pp. 93–114, Newbury Park, CA: Sage.

Clark, A. N., Manikkar, G. D. and Gray, I. (1975) Diogenes' syndrome: a clinical study of gross neglect in old age, *Lancet* i (7903), 366–8.

Clark, F., Carlson, M., Zemke, R., Frank, G., Patterson, K., Ennevor, B. L., Rankin-Martinez, A., Hobson, L., Crandall, J., Mandel, D. and Lipson, L. (1996) Life domains and adaptive strategies of a group of low-income, well older adults, *American Journal of Occupational Therapy*, 50: 99–108.

Clarke (1992) Public faces, private lives: August Sander and the social typology of the portrait photograph, in G. Clarke (ed.) *The Portrait in Photography*, Seattle: Reaktion Books.

Clarkson, P., Hughes, J. and Challis, D. (2003) Public funding for residential and nursing home care: projection of the potential impact of proposals to change the residential allowance in services for older people, *International Journal of Geriatric Psychiatry*, 18: 211–16.

Cliff, Dallas (1993) Health issues in early male retirement, in S. Platt, H. Thomas, S. Scott and G. Williams (eds) *Locating Health: Sociological and historical explanations*, pp. 121–37, Aldershot: Avebury.

Cloke, P. (1996) Rural life-styles: material opportunity, cultural experience and how theory can undermine policy, *Economic Geography*, 72: 433–49.

Cloke, P. and Le Heron, R. (1994) Agricultural deregulation: the case of New Zealand, in P. Lowe, T. Marsden and S. Whatmore (eds) *Regulating Agriculture*, 104–26, London: David Fulton.

Cloutier-Fisher, Denise and Joseph, Alun (2000) Long-term care restructuring in rural Ontario: retrieving community service user and provider narratives, *Social Science and Medicine* 50, 7–8: 1037–45.

Coale, A. J. (1964) How a population ages or grows younger, in R. Freedman (ed.) *Population: The vital revolution*, pp. 47–58, New York: Anchor Books.

Cohen, L. (1995) Toward an anthropology of senility: anger, weakness, and Alzheimer's in Banaras, India, *Medical Anthropology Quarterly*, 9, 3: 314–34.

Cohen, L. (1998) *No Aging in India: Alzheimer's, the bad family, and other modern things*, Berkeley: University of California Press.

Cohen, M. N. (1989) *Health and the Rise of Civilisation*, New Haven: Yale University Press.

Cohen, S. and Taylor, L. (1992) *Escape Attempts: The theory and practice of resistance to everyday life* (2nd edn), London: Routledge.

Cole, T. R. (1993) Preface, in T. Cole, W. Achenbaum, P. Jakobi and R. Kastenbaum (eds) *Voices and Visions of Aging: Toward a critical gerontology*, pp. vii–xi, New York: Springer Publishing.

Cole, T. R., Kastenbaum, R. and Ray, R. E. (eds) (2000) *Handbook of the Humanities and Aging*, 2nd edn, New York: Springer Publishing.

Coleman, P. (1993a) Adjustment in later life, in J. Bond, P. Coleman and S. Peace (eds) *Ageing in Society: An introduction to social gerontology* (2nd edn), pp. 97–132, London: Sage.

Coleman, P. (1993b) Psychological ageing, in J. Bond, P. Coleman and S. Peace (eds) *Ageing in Society: An introduction to social gerontology* (2nd edn), pp. 68–96, London: Sage.

Collins, D. and Kearns, R. A. (1998) *Avoiding the Log-Jam: Exotic forestry, transport and health in Hokianga*, Working Paper no. 8, Department of Geography, University of Auckland.

Commonwealth Department of Health and Ageing (2002) *National Strategy for an Ageing Australia*. Retrieved 21 January 2003 from http://www.olderaustralians.gov.au/documents/pdf/nsaabook.pdf.

Conway, S. (2000) 'I'd be unhealthy if nobody wanted me any more': a sociological analysis of the relationship between ageing and health beliefs, unpublished Ph.D. thesis, University of Hull.

Conway, S. and Hockey, J. (1998) Resisting the mask of old age: the social meaning of lay health beliefs in later life, *Ageing and Society*, 18: 469–94.

Costa, P. T. and McCrae, R. R. (1988) Personality in adulthood: a six-year longitudinal study of self-reports and spouse ratings on the NEO Personality Inventory, *Journal of Personality and Social Psychology*, 54: 853–63.

Costa, P. T. and McCrae, R. R. (1992) *Professional Manual: Revised NEO Personality Inventory (NEO PI-R) and NEO Five-Factor Inventory (NEOFFI)*, Odessa, FL: Psychological Assessment Resources.

Costa, P. T. and McCrae, R. R. (1994) Set like plaster? Evidence for the stability of adult personality, in T. F. Heatherton and J. L. Weinberger (eds) *Can Personality Change?* pp. 21–40, Washington, DC: American Psychological Association.

Courts, Nancy Fleming, Barba, Beth E. and Tesh, Anita (2001) Family caregivers' attitudes toward aging, caregiving, and nursing home placement, *Journal of Gerontological Nursing*, 27, 8: 44–52.

Cowgill, D. O. and Holmes, L. D. (1970) The demography of ageing, in A. M. Hoffman (ed.) *The Daily Needs and Interests of Older People*, pp. 27–69, Springfield, Illinois: Charles C. Thomas.

Coyte, Peter C. and McKeever, Patricia (2001) Home care in Canada: passing the buck, *Canadian Journal of Nursing Research*, 33, 2: 11–25.

Creighton Campbell, John and Ikegami, Naoki (2003) Japan's radical reform of long-term care, *Social Policy and Administration*, 37, 1: 21–34.

Cumming, E. and Henry, W. (1961) *Growing Old: The process of disengagement*, New York: Basic Books.

Cunningham, I. (1977) *After Ninety*, Seattle: University of Washington Press.

Curtice, Lisa and Fraser, Fiona (2000) The domiciliary care market in Scotland: quasi-markets revisited, *Health and Social Care in the Community*, 8, 4: 260–8.

Cutchin, M. (2003) The process of mediated aging-in-place: a theoretically and empirically based model, *Social Science and Medicine*, 57: 1,077–90.
Dahms, F. A. (1996) The greying of south Georgian Bay, *the Canadian Geographer*, 40: 248–163.
Dannefer, D. and Uhlenberg, P. (1999) Paths of the life course: a typology, in V. L. Bengston and K. W. Schaie (eds) *Handbook of Theories of Aging*, pp. 306–26, New York: Springer.
Dante, A. (1949–69; orig. 1308), *Divina Commedia in English* (Inferno – Prologue – Cantos 1 and 2), trans. D. L. Sayers, Harmondsworth: Penguin.
Dargent-Molina, P., Hays, M. and Breart, G. (1996) Sensory impairments and physical disability in aged women living at home, *International Journal of Epedemiology*, 25: 621–64.
Darton, R., Netten, A. and Forder, J. (2003) The cost implications of the changing population and characteristics of care homes, *International Journal of Geriatric Psychiatry*, 18: 236–43.
Davies, J. (2001) Partnership working in health promotion: the potential role of social capital in health development, in S. Balloch and M. Taylor (eds) *Partnership Working: Policy and practice*, Bristol: The Policy Press.
Davis, B. and Knapp, M. (1981) *Old People's Homes and the Production of Welfare*, London: Routledge.
Day, A. T. (1991) *Remarkable Survivors: Insights into successful aging among women*, Washington, DC: Urban Institute Press.
Day, L., Kent, S. and Fildes, B. (1994) Injuries among older people, *Hazard*, 19: 1–16.
Day, P., Klein, R. and Redmayne, S. (1997) *Why Regulate? Regulating residential care for elderly people*, Bristol: Rowntree.
De Raedt, R. and Ponjaert-Kristoffersen, I. (2000) Can strategic and tactical compensation reduce crash risk in older drivers? *Age and Ageing*, 29: 517–21.
Dear, M. J. and Wolch, J. (1987) *Landscapes of Despair*, Oxford: Polity Press.
Dear, M. J. and Wolch, J. (1989) *Power in Geography: How territory shapes social life*, Boston, MA: Unwin Hyman.
Department of Health (1989) *Homes Are for Living in*, London: HMSO.
Department of Health (1999) *Fit for the Future? National required standards for residential and nursing homes for older people*, London: HMSO.
Department of Health (2001) *Care Homes for Older People: National minimum standards* London: HMSO.
Department of Health and Social Security (1984) *Home Life*, London: Centre for Policy on Ageing.
Diamond, T. (1992) *Making Gray Gold: Narratives of nursing home care*, Chicago: University of Chicago Press.
Diehl, M., Coyle, N. and Labouvie-Vief, G. (1996) Age and sex differences in strategies of coping and defense across the lifespan, *Psychology and Aging*, 11: 127–39.
Domosh, M. (1998) Geography and gender: home, again? *Progress in Human Geography*, 22, 2: 276–82.
Donaghy, Martin (1999) Commentary (on Zarit *et. al.*, 1999 'Useful services for families: research findings and directions'), *International Journal of Geriatric Psychiatry*, 14, 3: 165–81.
Dorn, M. L. and Laws, G. (1994) Social theory, body politics and medical geography, *Professional Geographer*, 46: 106–10.
Doty, Pamela, Jackson, Mary E. and Crown, William (1998) The impact of female

caregivers' employment status on patterns of formal and informal eldercare, *The Gerontologist*, 38, 3: 4–14.

Douglas, D. (1999) The new rural regions: consciousness, collaboration and new challenges and opportunities for innovative practice, in W. Ramp, J. Kulig, L. Townshend and V. McGowan (eds) *Health in Rural Settings: Contexts for action*, pp. 39–60, Alberta: University of Lethbridge.

Douglas, M. (1986) *How Institutions Think*, Syracuse, NY: Syracuse University Press.

Du Plessis, V., Beshiri, R. and Bollman, R. D. (2001) Definitions of rural, *Rural and Small Town Canada Bulletin*, 3, Ottawa: Statistics Canada, (Catalogue 21–006–X1E).

Duffy, M., Bailey, S., Beck, B. and Barker, D. G. (1996) P in nursing home design: a comparison of residents, administrators and designers, *Environment and Behaviour*, 18, 2: 246–57.

Duncan, J. (1990) *The City as Text: The politics of landscape interpretation in the Kandyan kingdom*, Cambridge: Cambridge University Press.

Duncan, J. and Duncan, N. (1988) (Re)reading the landscape, *Environment and Planning D: Society and Space*, 6: 117–26.

Dunkle, R. E., Kart, C. S. and Lockery, S. A. (1994) Self-care, in B. R. Bonder and M. B. Wagner (eds) *Functional Performance in Older Adults*, pp. 122–35, Philadelphia: F. A. Davis Company.

Dyer, I. (1979) The role of stereotypes, in J. Cook and M. Lewington (eds) *Images of Alcoholism*, London: BFI.

Eales, J., Keating, N. and Damsma, A. (2003) Seniors' experiences of client-centred residential care, *Ageing and Society*, 21 279–96.

Earhart, C. C. and M. J. Weber (1992) Mobility intentions of rural elderly and non-elderly: implication for social services and planned housing for the elderly, *Journal of Housing for the Elderly* 10, 1–2: 49–64.

Earle, L. (1996) *Successful Ageing in Australian Society: A community development challenge*, Adelaide: Recreation for Older Adults.

Ebrahim, S. and Kalache, A. (eds) (1996) *Epidemiology in Old Age*, London: British Medical Association Publishing Group.

Eckert, J. K., Cox, D. and Morgan, L. (1999) The meaning of family-like care among operators of small board care homes, *Journal of Aging Studies*, 13, 3: 333–47.

Economic Commission for Asia and the Pacific (UN ESCAP) (2002) National policies and programmes on ageing in Asia and the Pacific: an overview and lessons learned, *Social Policy Paper No. 9*, New York: United Nations.

Edelman, Perry and Hughes, Susan (1990) The impact of community care on the provision of informal care to home-bound elderly persons, *Journal of Gerontology* 45: S74–S84.

Elder, G. (1977) *The Alienated: Growing old today*, London: Writers' and Readers' Co-operative.

Elley, C. R., Kerse, N., Aroll, B. and Robinson, E. (2003) Effectiveness of counselling patients on physical activity in general practice: cluster randomised controlled trial, *BMJ*, 326: 1–6.

Emmison, M. and Smith, P. (2000) *Researching the Visual*, London: Sage.

England, Kim (2000) 'It's really hitting home': the home as a site for long-term health care, *International Women and Environments Magazine*, summer/fall (48/49): 25.

England, Suzanne E. and Linsk, Nathan L. (1989) Paid to care for their own: a report on a community care program that permits relatives to be hired as caregivers, *Home Health Care Services Quarterly* 10, 1–2: 61–71.

England, Suzanne E., Linsk, Nathan L., Simon-Rusinowitz, Lori and Keigher, Sharon M. (1990) Paying kin for care: agency barriers to formalizing informal care, *Journal of Aging and Social Policy*, 2, 2: 63–86.

Erikson, E. E. (1959/1980) *Identity and the Life Cycle*, New York: Norton.

Erikson, E. E., Erikson, J. M. and Kivnick, H. Q. (1986) *Vital Involvement in Old Age*, New York: W. W. Norton and Company.

European Institute for Design and Disability (EIDD) (2004) http://www.design-for-all.org/.EIDD. Website accessed 25 August 2004.

Evans, B. (1993) Health care reform: the issue from hell, *Policy Options*, 14: 35–41.

Evans, G. W. (1999) Measurement of the physical environment as stressor, in S. L. Friedman and T. D. Wachs (eds) *Measuring Environment across the Life Span: Emerging methods and concepts*, pp. 249–78, Washington, DC: American Psychological Association.

Evans, G. W. and Cohen, S. (1987) Environmental stress, in D. Stokols and I. Altman (eds) *Handbook of Environmental Psychology* (vol. 1), New York: John Wiley.

Everitt, J. (1994) A tale of two towns: the geography of aging in southern Manitoba, *Small Town*, 24: 4–11.

Everitt, J. and Gfellner, B. (1996) Elderly mobility in a rural area: the example of southwest Manitoba, *Canadian Geographer*, 40: 338–51.

Fabian, J. (1993) Crossing and patrolling: thoughts on anthropology and boundaries, *Culture*, 13, 1: 49–53.

Fazel, S., Hope, T., O'Donnell, I., Piper, M. and Jacoby, R. (2001) Health of elderly male prisoners: worse than the general population, worse than younger prisoners, *Age and Ageing*, 30: 403–7.

Featherstone, M. (1991) *Consumer Culture and Postmodernism*, London: Sage.

Featherstone, M. (1995) Post-bodies, aging and virtual reality, in M. Featherstone and A. Wernick (eds) *Images of Aging: Cultural representations of later life*, pp. 227–44, London: Routledge.

Featherstone, M. and Hepworth, M. (1990) Images of ageing, in J. Bond and P. Coleman (eds) *Ageing in Society*, London: Sage.

Featherstone, M. and Hepworth, M. (1991) The mask of ageing and the postmodern life course, in M. Featherstone, M. Hepworth and B. S. Turner (eds) *The Body: Social process and cultural theory*, pp. 371–89, London: Sage.

Featherstone, M. and Wernick, A. (1995) *Images of Aging: Cultural representations of later life*, London: Routledge.

Fellegi, I. P. (1996) *Understanding Rural Canada: Structures and trends* (Statistics Canada: www.statcan.ca).

Finch, Janet (1989) *Family Obligations and Social Change*, Cambridge: Polity Press.

Finch, Janet and Groves, Dulcie (1982) By women, for women: caring for the frail elderly, *Women's Studies International Forum*, 5, 5: 427–38.

Finch, J. and Mason, J. (1993) *Negotiating Family Responsibilities*, London: Routledge.

Findlay, R. A. (2003) Interventions to reduce social isolation among older people: where is the evidence?, *Ageing and Society*, 23, 5: 647–58.

Fisher, J. and Giloth, R. (1999) Adapting rowhomes for Aging in Place: the story of Baltimore's 'Our Idea House', *Journal of Housing for the Elderly*, 13: 1–2: 3–18.

Fisk, M. (1999) *Our Future Home: Housing and the inclusion of older people in 2025*, London: Help the Aged.

Fiske, G. W. (1912) *The Challenge of the Country*, New York: Association Press.

Fisker, C. E. (2003) An Exploration of Aging in Rural Communities; The Potential of the Internet as a Means of Enhancing Independence, unpublished MA thesis, Department of Geography, University of Guelph, Ontario, Canada.

Flaskerud, J. H. and Winslow, B. J. (1998) Conceptualizing vulnerable populations: health-related research, *Nursing Research*, 47: 69–78.

Folkman, S. and Lazarus, R. S. (1990) Coping and emotion, in N. L. Stein, B. Leventhal and T. Trabasso (eds) *Psychological and Biological Approaches to Emotion*, Hillsdale, NJ: Erlbaum.

Fonda, S. J., Wallace, R. B. and Herzog, A. R. (2001) Changes in driving patterns and worsening depressive symptoms among older adults, *Journal of Gerontology* 56B, 6: S343–S351.

Fontana, A. and Smith, R. W. (1989) Alzheimer's disease victims: the 'unbecoming' of self and the normalization of competence, *Sociological Perspectives*, 32, 1: 35–46.

Ford, J. and Sinclair, R. (1987) *Sixty Years On: Women talk about old age*, London: Women's Press: 51.

Ford, R. G. and Smith, G. C. (1995) Spatial and structural change in institutional care for the elderly in south-east England, 1987–1990 *Environment and Planning A* 27: 225–48.

Forder, J. (1997) Contracts and purchaser–provider relationships in community care, *Journal of Health Economics*, 16: 517–42.

Forder, J. and Netten, A. (2000) The price of placements in residential and nursing home care: the effects of contracts and competition, *Health Economics*, 9: 643–57.

Foremen, R., Brookes, L., Abernethy, P., Brown, W. and Stoneham, M. (2001) Assessing the sustainability of the 'Just Walk It' program model: is it effective and will it enhance program success? Paper presented at Australia: Walking the 21st Century, Perth, WA. 20–2 February 2001.

Fox, N. (1999) Power, control and resistance in the timing of health and care, *Social Science and Medicine*, 48: 1307–19.

Frain, J. P. and Carr, P. H. (1996) Is the typical modern house designed for future adaptation for disabled older people? *Age and Ageing*, 25: 398–401.

Fratiglioni, L., Wang, H.-X., Ericsson, K., Maytan, M. and Winblad, B. (2000) Influence of social network on occurrence of dementia: a community-based longitudinal study, *The Lancet*, 355, 9212: 1315–20.

Frederiks, Carla M. A., te Wierik, Magreet J. M., Visser, Adriaan and Sturmans, Ferd (1990) The functional status and utilisation of care by elderly people living at home, *Journal of Community Health*, 15, 5: 307–17.

'Free Radicals' (1977) *Generations Review*, 7: 24–5.

French, P. (2003) Neither shy nor retiring, *Observer Review*, 26 January: 7.

Fries, J. F. (1980) Aging, natural death, and the compression of morbidity, *New England Journal of Medicine*, 303, 3: 130–5.

Fries, J. F. (1996) Physical activity, the compression of morbidity, and the health of the elderly, *J R Soc Med*, 89, 2: 64–8.

Fuller, A. M. (1994) Sustainable rural communities in the arena society, in J. M. Brydon (ed.) *Towards Sustainable Rural Communities: The Guelph Seminar Series*, pp. 133–9, Ontario, Canada: the University of Guelph.

Fuller, A. M. (1997) Rural institutions in the arena society, in R. C. Rounds (ed.) *Changing Rural Institutions: A Canadian perspective*, pp. 41–52, Brandon, Canada: Canadian Rural Restructuring Foundation and the Rural Development Institute,.

Garfein, A. J. and Herzog, A. R. (1995) Robust aging among the young-old, old-old, and oldest-old, *The Journals of Gerontology. Series B, Psychological Sciences and Social Sciences*, 50, 2: S77–87.

Garner, E., Kempton, A. and van Beurden, E. (1996) Strategies to prevent falls: the Stay On Your Feet Program, *Australian Journal of Health Promotion*, 6: 337–43.

Gazerro, M., Inelmen, E., Secco, G. and Gatto, M. (1996) Elderly people: state of health and living environment. The case of Budrio (northern Italy), *Health and Place*, 2: 115–23.

Gesler, W. (1992) Therapeutic landscapes: medical issues in light of the new cultural geography, *Social Science and Medicine*, 34: 735–46.

Giddens, A. (1984) *The Constitution of Society*, Cambridge: Polity Press.

Giddens, A. (1991) *Modernity and Self-Identity*, Cambridge: Polity Press.

Gitterman, A. (1991) Introduction: social work practice with vulnerable populations, in A. Gitterman (ed.) *Handbook of Social Work Practice with Vulnerable Populations*, pp. 1–12, New York: Columbia University Press.

Glaser, B. and Strauss, A. L. (1971) *Status Passage*, London: Routledge and Kegan Paul.

Glasgow, N. and Blakely, R. M. (2000) Older non-metropolitan residents' evaluations of their transportation arrangements, *The Journal of Applied Gerontology*, 19, 1: 95–116.

Glass, D. C. and Singer, J. E. (1972) *Urban Stress*, New York: Academic Press.

Glazer, Nona Y. (1990) The home as a workshop: women as amateur nurses and medical care providers, *Gender and Society*, 4, 4: 479–99.

Glendinning, Caroline (1992) Employment and 'community care': policies for the 1990s, *Work, Employment and Society*, 6, 1: 103–11.

Glendinning, Caroline (ed.) (1998) *Rights and Realities: Comparing new developments in long-term care for older people*, London: The Policy Press.

Glendinning, Caroline; Schunk, Michaela and McLaughlin, Eithne (1997) Paying for long-term domiciliary care: a comparitive perspective, *Ageing and Society* 17: 123–40.

Glendinning, Caroline; Halliwell, Shirley; Jacobs, Sally; Rummery, Kirstein and Tyrer, Jane (2000) New kinds of care, new kinds of relationships: how purchasing services affects relationships in giving and receiving personal assistance, *Health and Social Care in the Community*, 8, 3: 201–11.

Glendinning, Caroline; Halliwell, Shirley; Jacobs, Sally; Rummery, Kirstein and Tyrer, Jane (2001) Bridging the gap: using direct payments to purchase integrated care, *Health and Social Care in the Community*, 8, 3: 192–200.

Golant, S. M. (1972) *The Residential Location and Spatial Behavior of the Elderly*, Chicago: University of Chicago Press.

Golant, S. M. (1984) *A Place to Grow Old: The meaning of environment in old age*, New York: Columbia University Press.

Golant, S. M. (1990) The metropolitanisation and suburbanisation of the U.S. elderly population: 1970–1988, *The Gerontologist*, 30: 80–5.

Goldberg, L. R. (1993) The structure of phenotypic personality traits, *American Psychologist*, 48: 26–34.

Golledge, R. G. and Stimson, R. J. (1997) *Spatial Behavior: a geographic perspective*, New York: Guilford Press.

Gould, S. J. (1999) A critique of Heckhausen and Schulz's (1995) lifespan theory of control from a cross-cultural perspective, *Psychological Review*, 106, 3: 597–604.

Goward, L. (1995) *At the Public's Convenience: A report on progress towards public access to reports on inspection visits to residential care homes for older people in Greater London*, London: Counsel and Care.

Graham, Hilary (1983) Caring: a labour of love, in J. Finch and D. Groves (eds) *A Labour of Love: Women, work, and caring*, London: Routledge and Kegan Paul: 131–42.

Graham, Hilary (1985) Providers, negotiators, and mediators: women as the hidden carers, in E. Lewin and V. Olesen (eds) *Women, Health, and Healing: Toward a new perspective*, pp. 25–52, New York: Tavistock.

Greene, Vernon (1983) Substitution between formally and informally provided care for the impaired elderly in the community, *Medical Care*, 21: 609–19.

Greenwell, L. and Bengston, V. (1997) Geographic distance and the contact between middle-aged children and their parents, *Journal of Gerontology*, 52B: 13–26.

Greenwood, Robert (2002) Aging in place: what do people want? Comments from PACE focus groups on long-term care, *Nursing Homes: Long-term care management*, 51, 2: 26–30.

Gribble, J. N. and Preston, S. H. (1993) *The Epidemiological Transition: Policy and planning implications for developing countries*, Washington, DC: National Academy Press.

Griffiths, P., Harris, R., Richardson, G., Hallett, N., Heard, S. and Wilson-Barnett, J. (2001) Substitution of a nursing led inpatient unit for acute services: randomised controlled trial of outcomes and cost-nurse-led intermediate care, *Age and Ageing*, 30: 483–8.

Groger, L. (1995) A nursing home can be a home, *Journal of Aging Studies*, 9, 2: 137–53.

Gross, B. and Caiden, M. (2000 Spring) The implications of aging in place for community-based services for elderly people, *Journal of Case Management: The Journal of Long-Term Home Health Care*, 2, 1: 21–6.

Grundy, E. and Bowling, A. (1999) Enhancing the quality of extended life years: identification of the oldest old with a very good and very poor quality of life, *Aging and Mental Health*, 3, 3: 199–212.

Gubrium, J. (1975) *Living and Dying at Murray Manor*, New York: St. Martin's.

Gubrium, J. (1986) *Oldtimers and Alzheimer's: The descriptive organization of senility*, Greenwich, CT: JAI Press.

Gubrium, J. (1987) Structuring and destructuring the course of illness: the Alzheimer's disease experience, *Sociology of Health and Illness*, 9, 1: 1–24.

Gubrium, J. (1993a) *Speaking of Life: Horizons of meaning for nursing home residents*, New York: Aldine DeGruyter.

Gubrium, J. (1993b) Voice and context in a new gerontology, in T. Cole, W. A. Achenbaum, P. L. Jakobi and R. Kastenbaum (eds) *Voices and Visions of Aging: Toward a critical gerontology*, pp. 46–64, New York: Springer Publishing.

Gubrium, J. and Holstein, J. A. (1999) The nursing home as a discursive anchor for the ageing body, *Ageing and Society*, 19: 519–38.

Gubrium, J. and Holstein, J. A. (2000) *Aging and Everyday Life*, Oxford: Blackwell Publishers.

Guest, R. S. and McDonald, I. M. (2002) Would a decrease in fertility be a threat to living standards in Australia?, *The Australian Economic Review*, 35, 1: 29–44.

Gullette, M. (2000) Age studies as cultural studies, in T. Cole, R. Kastenbaum and R. Ray (eds) *Handbook of the Humanities and Aging*, 2nd edn, pp. 214–34, New York: Springer Publishing Company.

Gurney, C. and Means, R. (1993) The meaning of home in later life, in S. Arber and M. Evandrou (eds) *Ageing, Independence and the Life Course*, London: Jessica Kingsley.

Hallman, Bonnie C. and Joseph, Alun E. (1997) Exploring distance and caregiver gender effects in eldercare: towards a geography of family caregiving, *Great Lakes Geographer*, 4, 2: 15–29.

Hallman, Bonnie C. and Joseph, Alun E. (1999) Getting there: mapping the gendered geography of caregiving to elderly relatives, *Canadian Journal on Aging*, 18, 4: 397–414.

Hamnett, C. and Mullings, B. (1992) The distribution of public and private residential homes for elderly persons in England and Wales, *Area*, 2, 2: 130–44.

Hancock, R. (1998) Housing wealth, income and financial wealth of older people in Britain, *Ageing and Society*, 18: 5–33.

Hanger, H. C., Ball, M. C. and Wood, L. A. (1999) An analysis of falls in the hospital: can we do without bed rails? *Journal of the American Geriatric Society*, 47: 529–31.

Hanlon, N. (2001) Hospital restructuring in smaller urban Ontario settings: unwritten rules and uncertain relations, *Canadian Geographer*, 45: 252–67.

Hanson, J., Kellaher, L. and Karmona, S. (forthcoming) *Re-defining Domesticity*, London: Housing Corporation.

Harburg, E., Erfrut, J. C., Chaperi, C., Hauenstein, L. S., Schull, W. J. and Schork, M. A. (1973) Socioecological stressor areas and black–white blood pressure, *Journal of Chronic Disease*, 26: 595–611.

Hareven, Tamara K. (1991a) The history of the family and the complexity of social change, *American History Review*, 96, 1: 95–124.

Hareven, Tamara K. (1977) Family time and historical time, *Daedalus*, 106: 57–70.

Hareven, Tamara K. (1991b) The home and the family in historical perspective, *Social Research*, 58, 1: 253–85.

Harper, R. D. and Dickson, W. A. (1995) Reducing the burn risk to elderly persons living in residential care, *Burns*, 21, 3: 205–8.

Harper, S. and Laws, G. (1995) Rethinking the geography of ageing, *Progress in Human Geography*, 19: 199–221.

Harris, Phyllis Brady (1993) The misunderstood caregiver? A qualitative study of the male caregiver of Alzheimer's disease victims, *The Gerontologist*, 33, 4: 551–56.

Harris, Phyllis Brady (1998) Listening to caregiving sons: misunderstood realities, *The Gerontologist*, 38, 3: 342–52.

Harris, Phyllis Brady and Long, Susan Orpett (1999) Husbands and sons in the United States and Japan: cultural expectations and caregiving experiences, *Journal of Aging Studies*, 13, 3: 241–67.

Harrop, A. and Grundy, M. D. (1991) Geographical variations in moves into institutions among elderly in England and Wales, *Urban Studies*, 28, 1: 65–86.

Hauxwell, H. (1989) *Seasons of My Life: The story of a solitary daleswoman*, London: Century.

Havinghurst, R. J. (1963) Successful aging, in R. Williams, C. Tibbits and W. Donahue (eds) *Process of Aging*, pp. 299–320, New York: Atherton.

Hazan, H. (1983) Discontinuity and identity, *Research on Aging*, 5, 4: 473–89.

Hazan, H. (1994) *Old Age: Constructions and deconstructions*, Cambridge: Cambridge University Press.

Hazan, H. (1996) *From First Principles: An experiment in ageing*, Westport, CN: Bergin and Garvey.

Health Canada (1997) The future of caregiving, *Seniors Info Exchange*, Winter 1997–8: 1–6.

Health Canada (1998) *International Activities in Tele-Homecare – background paper*, Toronto: Office of Health and the Information Highway, Health Canada.

Health Canada (1999) *Provincial and Territorial Home Care Programs: A synthesis for Canada*, Ottawa: Health Canada.

Health Canada (2001a) *Public Home Care Expenditures in Canada, 1975–76 to 1997–98*, Ottawa: Health System and Policy Division, Policy and Consultation Branch, Health Canada.

Health Canada (2001b) *Workshop on Healthy Aging*. Retrieved 23 May 2002 from www.hc-sc.gc.ca/seniors-aines/pubs/healthy_aging/intro_e.htm.

Health Canada (2002) National Profile of Family Caregivers in Canada – 2002: Final report, Ottawa: prepared for Health Canada by Decima Research Inc.

Health Transition Fund (1998) *Proceedings of the National Conference on Home Care*, Halifax:

Healthy Ageing Task Force (2000) *Commonwealth, State and Territory Strategy on Healthy Ageing*, Canberra: Commonwealth Department of Health and Aged Care.

Heckhausen, J. and Baltes, P. B. (1991) Perceived controllability of expected psychological change across adulthood and old age, *Journal of Gerontology: Psychological Sciences*, 46: 165–73.

Heckhausen, J. and Schulz, R. (1995) A lifespan theory of control, *Psychological Review*, 102: 284–304.

Helmchen, H., Baltes, M. M., Geiselmann, B., Kanowski, S., Linden, M., Reischies, F. M., Wagner, M., Wernicke, T. and Wilms, H. U. (1999) Psychiatric illness in old age, in P. B. Baltes and K. U. Meyer (eds) *The Berlin Aging Study: Aging from 70–100*, pp. 167–96, New York: Cambridge University Press.

Henderson, N. (1994) Bed, body and soul: the job of the nursing home aide, *Generations*, 18, 3: 20–2.

Henderson, N. and Vesperi, M. (eds) (1995) *The Culture of Long Term Care: Nursing home ethnography*, Westport, CT: Bergin and Garvey.

Hepworth, M. (2000) *Stories of Ageing*, Buckingham: Open University Press.

Hermanova, H. (1995) Healthy aging in Europe in the 1990s and implications for education and training in the care of the elderly, *Educational Gerontology*, 21, 1: 1–14.

Hevey, D. (1992) *The Creatures Time Forgot: photography and disability imagery*, London: Penguin.

Heywood, F., Pate, A., Galvin, J. and Means, R. (1999) *Housing Options for Older People*, London: Housing Corporation.

Hidalgo, M. C. and Hernandez, B. (2001) Place attachment: conceptual and empirical questions, *Journal of Environmental Psychology*, 21: 273–81.

Higgs, P., MacDonald, L., MacDonald, J. and Ward, M. (1998) Home from home: residents' opinions of nursing homes and long-stay wards, *Age and Ageing*, 27: 199–205.

Hill, A. J., Germa, F. and Boyle, J. C. (2002) Burns in older people: outcomes and risk factors, *Journal of the American Geriatric Society*, 50: 1912–13.

Hobbs, F. B. with Damon, B. L. (1996) *65+ in the United States*, US Bureau of the Census Current Population Reports, Special Studies P23–190, Washington, DC: US Government Printing Office.

Hockey, J. (1999) The ideal of home: domesticating the institutional space of old age and death, in T. Chapman and J. Hockey (eds) *Ideal Homes? Social Change and Domestic Life*, London: Routledge.

Hockey, J. and James, A. (1993) *Growing Up and Growing Old: Ageing and dependency in the life course*, London: Sage.

Hockey, J., Penhale, B. and Sibley, D. (2001) Landscapes of loss: spaces of memory, times of bereavement, *Ageing and Society*, 21: 739–57.

Hodge, G. (1991) *Seniors in Small Town British Columbia: Demographic tendencies and trends, 1961 to 1986*, Burnaby, British Columbia, Canada: Gerontology Research Centre, Simon Fraser University.

Hodge, G. (1993) *Canada's Aging Rural Population: The role and response of local government*, Toronto: ICURR Press.

Holahan, C. K. and Holahan, C. J. (1987) Life stress, hassles, and self-efficacy in aging: a replication and extension, *Journal of Applied Social Psychology*, 17: 574–92.

Holden, C. (2002) British government policy and the concentration of ownership in long-term care provision, *Ageing and Society*, 22: 70–94.

Holdsworth, D. W. and Laws, G. (1994) Landscapes of old age in coastal British Columbia, *Canadian Geographer*, 38: 174–81.

Holland, C. (2001) Housing histories: the experience of older women across the life course, unpublished Ph.D. thesis, Open University.

Hong Kong Government (2003) *SARS in Hong Kong: from Experience to Action: Report of the SARS Expert Committee* (October). Hong Kong: Government of the Hong Kong Special Administrative Region.

Hooyman, N. R. and Kiyak, H. A. (2002) *Social Gerontology: A multidisciplinary perspective*, Boston MA: Allyn and Bacon.

Horl, Josef (1993) Eldercare policy between the state and family: Austria, *Journal of Aging and Social Policy*, 5, 1–2: 155–68.

Hornstein, Z., Encel, S., Gunderson, M. and Neumark, S. (2001) *Outlawing Age Discrimination*, Bristol: The Policy Press.

Hoskins, J. (1998) *Biographical Objects: How things tell the stories of people's lives*, New York: Routledge.

Howden-Chapman, P., Signal, L. and Crane, J. (1999) Housing and health in older people: ageing in place, *Social Policy Journal of New Zealand* 13: 14–30.

Howell, S. (1980) *Design for Aging: Patterns of use*, Cambridge, MA: MIT Press.

Howell, S. C. (1983) The meaning of place in old age, in G. D. Rowles and R. J. Ohta (eds) *Aging and Milieu: Environmental perspectives in growing old*, pp. 97–107, New York: Academic Press.

Hubbard, G., Tester, S. and Downs, M. (2003) Meaningful social interactions between older people in institutional care settings, *Ageing and Society*, 23: 99–114.

Hugman R. (1999) Embodying old age, in E. Kenworthy-Teather (ed.) *Embodied Geographies: Spaces, bodies and rites of passage*, London: Routledge.

Hugman, R. (2001) Ageing in space, *Australian Journal on Ageing*, 20: 57–65.

Hunt, M. D. and Roll, M. K. (1987) Simulation in familiarizing older people with an unknown building, *The Gerontologist*, 27: 169–75.

Hutten, Jack B. F. and Kerkstra, Ada (eds) (1996) *Home Care in Europe: A country-specific guide to its organisation and financing*, Aldershot: Arena.

Ignatieff, M. (1993) *Scar Tissue*, London: Penguin Books.

Ignatieff, M. (11 July 2002) 'Interview with Eleanor Wachtel', *[The Arts Today]*, Toronto: Canadian Broadcasting Corporation.

Ilbery, B. (ed.) (1998) *The Geography of Rural Change*, London: Longmans.

Illiffe, S. and Lenihan, P. (2001) Promoting innovative primary care for older people in general practice using a community-oriented approach, *Primary Health Care Research and Development*, 2: 71–9.

Imrie, R. (1996) *Disability and the City: International perspectives*, New York: St Martin's Press.

Inouye, S. (1994) The dilemma of delirium: clinical and research controversies regarding diagnosis and evaluation of delirium in hospitalized elderly medical patients, *American Journal of Medicine*, 97: 278–88.

Irizarry, C., West, D. and Downing, A. (2001) Use of the Internet by older rural South Australians; *Australian Journal on Ageing*, 20: 2–34.

ITV (1997) *3D* documentary/magazine programme, 1 May.

Iwashyna, T. J. and Christakis, N. A. (2003) Marriage, widowhood, and health-care use, *Social Science and Medicine*, 57, 11: 2137–47.

Jackson, M. (1996) Introduction: phenomenology, radical empiricism, and anthropological critique, in M. Jackson (ed.) *Things as They Are: New directions in phenomenological anthropology*, pp. 1–50, Bloomington: Indiana University Press.

Jenkins, H. and Allen, C. (1998) The relationship between staff burnout/distress and interactions with residents in two residential homes for older people, *International Journal of Geriatric Psychiatry*, 13: 466–72.

Johansson, Lennarth (1991) Informal care of dependent elderly at home – some Swedish experiences, *Ageing and Society* 11: 41–58.

Johansson, Lennarth, Sundström, Gerdt and Hassing, Linda (2003) State provision down, offspring's up: the reverse substitution of old-age care in Sweden, *Ageing and Society*, 23, 3: 269–80.

John, O. P. (1990) The big five factor taxonomy: dimensions of personality in the natural language and in questionnaires, in L. Pervin (ed.) *Handbook of Personality: Theory and research*, pp. 66–100, New York: Guilford Press.

Johnson, J. (1998) The emergence of care as policy, in A. Brechin, J. Walmslely, J. Katz and S. Peace (eds) *Care Matters*, London: Sage.

Johnson, J. and Bytheway, B. (1997) Illustrating care: images of care relationships with older people, in A. Jamieson, *et al.* (eds) *Critical Approaches to Ageing and Later Life*, pp. 132–42, Buckingham: Open University Press.

Joint Interim Committee on Aging (1998) *Hearing*, Jefferson City, MO: [publisher?].

Jorm, A. F., Korten, A. E. and Henderson, A. S. (1987) The prevalence of dementia: a quantitative integration of the literature, *Acta Psychiatrica Scandinavica*, 76: 465–79.

Joseph, Alun E. (1996) Restructuring long-term care and the geography of aging: a view from rural New Zealand, *Social Science and Medicine*, 42: 887–96.

Joseph, Alun E. (1999) *Towards an Understanding of the Interrelated Dynamics of Change in Agriculture and Rural Community*, Population Studies Centre, University of Waikato, New Zealand, Discussion Paper 32.

Joseph, Alun E. (2002) Rural population and rural services, in I. R. Bowler, C. R. Bryant and C. Cocklin (eds) *The Sustainability of Rural Systems: Geographical interpretations*, pp. 211–24, London: Kluwer Academic Publishers.

Joseph, Alun E. and Bantock, P. R. (1984) Rural accessibility of general practitioners – the case of Bruce and Grey Counties, Ontario, 1901–1981, *Canadian Geographer*, 28: 226–39.

Joseph, Alun E. and Chalmers, A. I. (1995) Growing old in place: a view from rural New Zealand, *Health and Place*, 1: 79–90.

Joseph, Alun E. and Chalmers, A. I. (1998) Coping with rural change: finding a place for the elderly in sustainable communities, *New Zealand Geographer*, 52: 28–36.

Joseph, Alun E. and Chalmers, A. I. (1999) Residential and support services for the Waikato elderly, 1992–1997: privatisation and emerging resistance, *Social Policy Journal of New Zealand*, 13: 154–69.

Joseph, Alun E. and Cloutier, D. (1991) Elderly migration and its implications for service provision in rural communities: an Ontario perspective, *Journal of Rural Studies*, 7: 433–44.

Joseph, Alun E. and Fuller, A. M. (1991) Towards an integrative perspective on the housing, services and transportation implications of rural aging, *Canadian Journal on Aging*, 10: 127–48.

Joseph, Alun E. and Hallman, Bonnie C. (1996) Caught in the triangle: the influence of home, work, and elder location on work–family balance, *Canadian Journal on Aging*, 15, 3: 392–412.

Joseph, Alun E. and Hallman, Bonnie C. (1998) Over the hill and far away: distance as a barrier to the provision of assistance to elderly relatives, *Social Science and Medicine*, 46, 6: 631–9.

Joseph, Alun E. and Hollett, R. G. (1992) When I'm 65: the retirement housing preferences of the rural elderly, *Canadian Journal of Regional Science*, 15: 1–19.

Joseph, Alun E. and Martin-Matthews, A. (1993) Growing old in aging communities, *Journal of Canadian Studies*, 28: 14–29.

Joseph, Alun E. and Martin-Matthews, Anne (1994) Caring for elderly people: workforce issues and development questions, in D. Phillips and Y. Verhasselt (eds) *Health and Development*, pp. 161–81, London: Routledge.

Joseph, Alun E. and Phillips, D. R. (1999) Ageing in rural China: impacts of increasing diversity in family and community resources, *Journal of Cross-Cultural Gerontology*, 14: 153–68.

Joseph, Alun E., Lidgard, J. M. and Bedford, R. D. (2001) Dealing with ambiguity: on the interdependence of change in agriculture and rural communities, *New Zealand Geographer*, 57: 12–26.

Kahana, B. and Kahana, E. (1983) Stress reactions, in P. M. Lewinsohn and L. Teri (eds) *Clinical Geropsychology: New directions in assessment and treatment*, pp. 139–69, New York: Pergamon Press.

Kahana, E. and Kahana, B. (1982) A congruence model of person–environment interaction, in M. P. Lawton, P. G. Windley and T. O. Byerts (eds) *Aging and the Environment: Theoretical approaches*, pp. 97–121, New York: Springer.

Kahana, E., Kahana, B. and Kercher, K. (2003) Emerging lifestyles and proactive options on successful aging, *Ageing International*, 28, 2: 155–80.

Kalache, A. and Keller, I. (2000) The greying world: a challenge for the twenty-first century, *Science Progress*, 83, 1: 33–54.

Kane, Robert (1999) Examining the efficiency of home care, *Journal of Aging and Health*, 11, 3: 322–40.

Kart, C. S. and Kinney, J. M. (2001) *The Realities of Aging: An introduction to gerontology*, Boston, MA: Allyn and Bacon.

Kastenbaum, R. J. (1993) Encrusted elders: Arizona and the political spirit of postmodern aging, in T. R. Cole, A. Achenbaum, P. Jakobi and R. Kastenbaum (eds) *Voices and Visions of Aging: Toward a critical gerontology*, pp. 160–83, New York: Springer.

Kastenbaum, R. J. (1994) *Defining Acts: Aging as drama*, Amityville, NY: Baywood Publishing.

Katz, S. (1995) Imagining the life-span: from premodern miracles to postmodern fantasies, in M. Featherstone and A. Wernick (eds) *Images of Aging: Cultural representations of later life*, pp. 61–75, London: Routledge.

Katz, S. (1996) *Disciplining Old Age: The formation of gerontological knowledge*, Charlottesville: University Press of Virginia.

Katz, S. (2003) Critical gerontological theory: intellectual fieldwork and the nomadic life of ideas, in S. Biggs, A. Lowenstein and J. Hendricks (eds) *The Need for Theory: Critical approaches to social gerontology*, pp. 15–31, Amityville, NY: Baywood Publishing.

Katz, S. (forthcoming) *Cultural Aging: Essays on lifecourse, lifestyle and senior worlds*, Peterborough: Broadview Press.

Katz, S. and Marshall, B. (2003) New sex for old: lifestyle, consumerism, and the ethics of aging well, *Journal of Aging Studies*, 17: 3–16.

Kaufman, S. (1986) *The Ageless Self: Sources of meaning in later life*, Madison: University of Wisconsin Press.

Keane, W. L. (1994) The patient's perspective: the Alzheimer's association, *Alzheimer Disease and Associated Disorders*, 8, Suppl. 3: 151–5.

Kearns R. A. (1993) Place and health: towards a reformed medical geography, *The Professional Geographer*, 46: 67–72.

Kearns, R. A. and Joseph, A. E. (1993) Space in its place: developing a link in medical geography, *Social Science and Medicine*, 37: 711–17.

Kearns, R. A. and Joseph, A. E. (1997) Restructuring health and rural communities in New Zealand, *Progress in Human Geography*, 21: 18–32.

Kearns, R. A. and Moon, G. (2002) From medical to health geography: novelty, place and theory after a decade of change, *Progress in Human Geography*, 26: 605–25.

Keating, N., Keefe, J. and Dobbs, B. (2001) A good place to grow old? Rural communities and support to seniors, in R. Epp. and O. Whitson (eds) *Writing off the Rural West: Globalization, governments and the transformation of rural communities*, pp. 263–77, Edmonton: University of Alberta Press.

Keen, J. (1989) Interiors: architecture in the lives of people with dementia, *International Journal of Geriatric Psychiatry*, 4: 255–72.

Kellaher, L., Peace, S. and Willcocks, D. (1990) Triangulating data, in S. Peace (ed.) *Researching Social Gerontology*, pp. 115–28, London: Sage Publications.

Kelsey, J. (1997) *The New Zealand Experiment: A world model for structural adjustment?* (2nd edn), Auckland: Auckland University Press.

Kendig, H., Andrews, G., Browning, C., Quine, S. and Parsons, A. (2001) *A Review of Healthy Ageing Research in Australia*, Canberra, ACT: Commonwealth Department of Health and Aged Care.

Kendig, H., Helme, R., Teshuva, K., Osborne, D., Flicker, L. and Browning, C. (1996) *Health Status of Older People: Preliminary findings from a survey of the health and lifestyles of older Australians*, Melbourne: Victoria Health Promotion Foundation.

Kerschner, H. and Pegues, J. A. (1998) Productive aging: a quality of life agenda. *Journal of the American Dietetic Assocation*, 98, 12: 1445–8.

Kerse, N. M., Murphy, M. J., Flicker, L. and Young, D. (1997) Health promotion and older people: a qualitative study of general practitioners' views, *Medical Journal of Australia*, 167: 423–7.

Keyes, C. L. M. and Ryff, C. D. (1998) Generativity in adult lives: social structural contours and quality of life consequences, in D. P. McAdams and E. de St Aubins (eds) *Generativity and Adult Development: How and why we care for the next generation*, Washington, DC: American Psychological Association.

King, R., Warnes, T. and Williams, A. (2000) *Sunset Lives: British retirement migration to the Mediterranean*, Oxford: Berg.

Kinsella, K. and Phillips, D. R. (2005) *Global Aging Population Bulletin*, 1, November.

Kinsella, K. and Velkoff, V. A. (2001) *An Aging World: 2001*, US Census Bureau, Series P95/01–1, Washington, DC: US Government Printing Office.

Kitwood, T. (1990) The dialectics of dementia: with particular reference to Alzheimer's disease, *Ageing and Society*, 10: 177–96.

Kitwood, T. (1993) Towards a theory of dementia care: the interpersonal process, *Ageing and Society*, 13: 51–67.

Kitwood, T. (1997) *Dementia Reconsidered: The person comes first*, Buckingham: Open University Press.

Kitwood, T. (1998) Toward a theory of dementia care: ethics and interaction, *Journal of Clinical Ethics*, 9, 1: 23–34.

Kitwood, T. and Bredin, K. (1992) Towards a theory of dementia care: personhood and well-being, *Ageing and Society*, 12: 269–87.

Klein, J. T. (1996) *Crossing Boundaries: Knowledge, disciplinarities, and interdisciplinarities*, Charlottesville: University Press of Virginia.

Knijn, Trudie and Kremer, Monique (1997) Gender and the caring dimension of welfare states: towards an inclusive citizenship, *Social Politics*, Fall: 328–61.

Kohler, H.-P., Billari, F. C. and Ortega, J. A. (2002) The emergence of lowest-low fertility in Europe during the 1990s, *Population and Development Review*, 28, 4: 641–80.

Kolanowski, A. M. (1990) Restlessness in the elderly: the effect of artificial lighting, *Nursing Research*, 39, 3: 181–3.

Kontos, Pia C. (1998) Resisting institutionalization: constructing old age and negotiating home, *Journal of Aging Studies* 12: 167–84.

Koopman-Boyden. P. (ed.) (1993) *New Zealand's Ageing Society: The implications*, Wellington: Daphne Brassell Associates Press.

Koskinen, S. (2002) The village community as a resource for the aged, paper given at the BSG Annual Conference 2002, Active Ageing: Myth or Reality?, Birmingham.

KPMG (1998) *Banking and Finance: 1998 financial institutions performance survey*, Wellington: KPMG.

Krause, N. (1998) Neighborhood deterioration, religious coping, and changes in health during late life, *The Gerontologist*, 38: 653–64.

Krause, N. (2001) Social support, in R. H. Binstock and L. K. George (eds) *Handbook of Aging and the Social Sciences* (5th edn), pp. 272–94, San Diego, CA: Academic Press.

Krivo, Lauren J. and Mutchler, Jan E. (1989) Elderly persons living alone: the effect of community context on living arrangements, *Journal of Gerontology, Social Sciences*, 44, 2: 54–62.

Kuypers, J. A. and Bengston, V. L. (1973) Social breakdown and competence: a model of normal ageing, *Human Development*, 16: 181–201.

Labouvie-Vief, G., Diehl, M., Tarnowski, A. and Shen, J. (2000) Age differences in adult personality: findings from the United States and China, *The Journals of Gerontology: Psychological sciences and social sciences* 55B, 1: 4–17.

Labouvie-Vief, G., Hoyer, W. J., Baltes, M. M. and Baltes, P. B. (1974) Operant analysis of intellectual behaviour in old age, *Human Development*, 17: 259–72.

Lach, H. W., Reed, A. T., Arfken, C. L. *et al.* (1991) Falls in the elderly: reliability of a classification system, *Journal of the American Geriatric Society*, 39: 197–202.

Lachman, M. E. and Burack, O. R. (1993) Planning and control processes across the lifespan: an overview, *International Journal of Behavioral Development*, 16: 131–43.

Lai, C. K. and Arthur, D. G. (2003) Wandering behaviour in people with dementia, *Journal of Advanced Nursing*, 44: 173–82.

Laing, W. (1995) *Laing's Review of Private Healthcare 1995*, London: Laing and Buisson.

Lamb, S. E., Bartlett, H. P., Ashley, A. and Bird, W. (2002) Can lay-led walking programmes increase physical activity in middle-aged adults? A randomised controlled trial, *Journal of Epidemiology and Community Health*, 54, 4: 246–52.

Lang, F. R. and Baltes, M. M. (1997) Being with people and being alone in late life: costs and benefits for everyday functioning, *International Journal of Behavioral Development*, 21: 729–46.

Lang, F. R. and Carstensen, L. L. (1994) Close emotional relationships in later life: further support for proactive aging in the social domain, *Psychology and Aging*, 9: 315–24.

Lang, F. R. and Carstensen, L. L. (1998) Social relationships and adaptation in late life, in A. S. Bellack and M. Hersen (eds) *Comprehensive Clinical Psychology* (vol. 7), pp. 55–72, Oxford: Pergamon.

Lang, F. R., Ludtke, O. and Asendorpf, J. (2001) Validity and psychometric equivalence of the German version of the Big Five Inventory (BFI) in early, middle, and late adulthood, *Diagnostica*, 47: 111–21.

Langer, E. J. and Rodin, J. (1976) The effects of choice and enhanced personal responsibility for the aged: a field experiment in an institutional setting, *Journal of Personality and Social Psychology*, 34: 191–8.

Langer, E. and Saegert, S. (1977) Crowding and cognitive control, *Journal of Personality and Social Psychology*, 35: 175–82.

Larder, D., Day, P. and Klein, R. (1986) *Institutional Care for the Elderly*, Bath: Centre for the Analysis of Social Policy, University of Bath.

Larragy, Joseph F. (1993) Formal service provision and the care of the elderly at home in Ireland, *Journal of Cross Cultural Gerontology*, 8, 4: 361–74.

Laslett, P. (1989) *A Fresh Map of Life: The emergence of the third age*, London: Weidenfeld and Nicolson.

Law, C. M. and Warnes, A. M. (1976) The changing geography of the elderly in England and Wales, *Transactions of the Institute of British Geographers*, NS 1: 453–71.

Laws, G. (1993) 'The land of old age': society's changing attitudes to built environments for elderly people, *Annals of the Association of American Geographers*, 83: 672–93.

Laws, G. (1989) Privatization and dependency on the local welfare state, in J. Wolch and M. Dear (eds) *The Power of Geography: How territory shapes social life*, pp. 238–57, Boston, MA: Unwin Hyman.

Laws, G. (1994) Contested meanings, the built environment and aging in place, *Environment and Planning, A*, 26: 1787–802.

Laws, G. (1995) 'Embodiment and emplacement: identities, representation and landscape in Sun City retirement communities, *International Journal of Aging and Human Development*, 40, 4: 253–80.

Laws, G. (1997) Spatiality and age relations, in A. Jamieson, S. Harper and C. Victor (eds) *Critical Approaches to Ageing and Later Life*, Buckingham: Open University Press.

Laws, G. and Radford, J. (1998) Place, identity and disability: narratives of intellectually disabled people in Toronto, in R. A. Kearns and W. M. Gesler (eds) *Putting Health into Place: Landscape, identity and wellbeing*, pp. 77–101, Syracuse, NY: Syracuse University Press,.

Lawton, M. P. (1972) Assessing the competencies of older people, in D. Kent, R. Kastenbaum and S. Sherman (eds) *Research Planning and Action for the Elderly*, pp. 122–43, New York: Behavioral Publications.

Lawton, M. P. (1975) *Planning and Managing Housing for the Elderly*, New York: Wiley and Sons.

Lawton, M. P. (1983) Environment and other determinants of well-being in older people, *The Gerontologist*, 23, 4: 349–57.

Lawton, M. P. (1985) The elderly in context: perspectives from environmental psychology and gerontology, *Environment and Behavior*, 17: 501–19.

Lawton, M. P. (1990) Residential environment and self-directedness among older people, *American Psychologist*, 45, 5: 638–40.

Lawton, M. P. (1991) Functional status and aging well, *Generations*, 15, 1: 31–4.

Lawton, M. P. (1999) Environmental taxonomy: generalizations from research with older adults, in S. Friedman and T. Wachs (eds) *Measuring Environment across the Life Span*, Washington, DC: American Psychological Association, pp. 185–96.

Lawton, M. P. and Nahemow, L. (1973) Ecology and the aging process, in C. Eisdorfer and M. P. Lawton (eds) *The Psychology of Adult Development and Aging*, Washington, DC: American Psychological Association.

Lawton, M. P. and Simon, B. (1968) The ecology of social relationships in housing for the elderly, *The Gerontologist*, 8: 108–15.

Lawton, M. P., Brody, E. M. and Turner-Massey, P. (1978) Relationships of environmental factors to changes in well-being, *The Gerontologist*, 18: 133–7.

Lawton, M. P., Weisman, G., Sloane, P., Norris-Baker, C., Calkis, M. and Zimmerman, Z. (2000) A professional environment assessment procedure for special care units for elders with dementing illness and its relationship to thc 'Therapeutic Environment Screening Schedule', *Alzheimers Disease and Associated Disorders*, 14: 28–38.

Lazarus, R. S. and Folkman, S. (1984) Coping and adaptation, in W. D. Gentry (ed.) *Handbook of Behavioral Medicine*, pp. 282–325, New York: Guilford Press.

Le Heron, R. and Pawson, E. (eds) (1996) *Changing Places: New Zealand in the nineties*, Auckland: Longman Paul: 247–59.

Lee, R. D. (1994) The formal demography of population aging, transfers and the economic life cycle, in L. G. Martin and S. H. Preston (eds) *Demography of Aging*, pp. 8–49, Washington, DC: National Academy Press.

Lee, Yoon-Ro and Sung, Kyu Taik (1998) Cultural influences on caregiving burden: cases of Koreans and Americans, *International Journal of Aging and Human Development*, 46, 2: 125–41.

Leete, R. and Alam, I. (1999) Asia's demographic miracle: 50 years of unprecedented change, *Asia-Pacific Population Journal*, 14, 4: 9–20.

Lefebvre, H. (1991) *The Social Production of Space*, Oxford: Blackwell.

Leinonen, R., Heikkinen, E. and Jylha, M. (2001) Predictors of decline in self-assessments of health among older people: a 5-year longitudinal study, *Social Science and Medicine*, 52, 9: 1329–41.

Levy, B. and Langer, E. (1994) Aging free from negative stereotypes: successful memory in China and among the American deaf, *Journal of Personality and Social Psychology*, 66: 989–97.

Liebkind, K. (1992) Ethnic identity – challenging the boundaries of social psychology, in G. M. Breakwell (ed.) *Social Psychology of Identity and the Self Concept*, Surrey: Surrey University Press.

Lilley, J. M., Arie, T. and Chilvers, C. E. D. (1995) Accidents involving older people: a review of the literature, *Age and Ageing*, 24: 346–65.

Lin, G. and Rogerson, P. A. (1995) Elderly parents and the geographic availability of their children, *Research on Aging*, 17: 303–31.

Linn, M. W., Gurel, L. and Linn, B. S. (1977) Patient outcome as a measure of quality of nursing home care, *American Journal of Public Health*, 67: 337–44.

Lipman, A. R. and Slater, R. (1976) Status and spatial approbation in eight homes for old people, *The Gerontologist*, 25: 275–84.

Lipman, A. R., Slater, R. and Harris, H. (1979) The quality of verbal interaction in homes for old people, *Gerontology*, 25: 275–84.

Litwak, Eugene (1985) *Helping the Elderly: The complementary roles of informal networks and formal systems*, New York: Guildford Press.

Litwak, Eugene, Messeri, Peter and Silverstein, Merril (1990) The role of formal and informal groups in providing help to older people, *Marriage and Family Review* 15, 1–2: 171–93.

Lloyd, L. (2000) Dying in old age: promoting well-being at the end of life, *Mortality*, 5, 2: 171–88.

Lo, C. P. (1984) The geography of the elderly in a modernizing society: the Hong Kong case, *Singapore Journal of Tropical Geography*, 5: 1–20.

Longino, C. F. and Marshall, V. W. (1990) North American research on seasonal migration, *Ageing and Society*, 10: 229–35.

Loo, W. M. (2000) Living environment and residential satisfaction of older persons in Hong Kong: a comparative study between Sha Tin New Town and Wan Chai old urban areas, unpublished M.Phil. thesis, Lingnan University, Hong Kong.

Lorber, Judith (1990) From the Editor, *Gender and Society*, 4, 4: 445–50.

Lord, S. R. and Menz, H. B. (2000) Visual contributions to postural stability in older adults, *Gerontology*, 46: 306–10.

Lord, S. R. and Sinnett, P. F. (1986) Femoral neck fractures, admission, bed use, outcome and projections, *Medical Journal of Australia*, 145: 493–6.

Lord, S. R., Sherrington, C. and Menz, H. B. (2001) *Falls in Older People: Risk factors and strategies for prevention*, Cambridge: Cambridge University Press.

Low, J. A. and Chan, D. K. Y. (2002) Air travel in older people, *Age and Ageing*, 31: 17–22.

Lowe, M., Kerride, I., McPhee, J. and Fairfull-Smith, H. (2000) A question of competence, *Age and Ageing*, 29: 179–82.

Lowenthal, D. (1985) *The Past is a Foreign Country*, Cambridge: Cambridge University Press.

Lund, D. A., Caserta, M. S. and Dimond, M. F. (1993) The course of spousal bereavement in later life, in M. S. Stroebe, W. Stroebe and R. D. Hansson (eds) *Handbook of Bereavement: Theory, research and intervention*, pp. 240–54, New York: Cambridge University Press.

Lyman, K. (1988) Infantilization of elders: day care for Alzheimer's disease victims, *Research in the Sociology of Health Care*, 7: 71–103.

Lyons, Karen S. and Zarit, Steven H. (1999) Formal and informal support: the great divide, *International Journal of Geriatric Psychiatry*, 14, 3: 183–92.

McAdams, D. P. and de St Aubins, E. (1992) A theory of generativity and its assessment through self-report, behavioral acts, and narrative themes in autobiography, *Journal of Personality and Social Psychology*, 62: 1003–15.

McCracken, K. (1985) Disaggregating the elderly, *Australian Geographer*, 16, 3: 218–24.

McCracken, K. and Curson, P. (2000) In sickness and in health: Sydney past and present, in J. Connell (ed.) *Sydney: The emergence of a world city*, pp. 96–118 and 352–6, Melbourne: Oxford University Press.

McCrae, R. R. and Costa, P. T. (1990) *Personality in Adulthood*, New York: Guilford Press.

McCrae, R. R., Costa, P. T., Pedrosa de Lima, M., Simtes, A., Ostendorf, E., Angleitner, A., Maruai, I., Bratko, D., Caprara, G. V., Barbaranelli, C., Chae, J.-H. and Pliedmont, R. L. (1999) Age differences in personality across the lifespan: parallels in five cultures, *Developmental Psychology*, 35: 466–77.

McCraig, M. J. and Frank, G. (1991) The able self: adaptive patterns and choices in independent living for a person with cerebral palsy, *American Journal of Occupational Therapy*, 45: 224–34.

McDaniel, Susan A. (2000) Untangling love and domination: challenges of home care for the elderly in a reconstructing Canada, *Journal of Canadian Studies*, 43, 2: 193–210.

McDaniel, Susan A. and Lewis, Robert (1997) Did they or didn't they: intergenerational supports in Canada's past and a case study of Brigus, Newfoundland, 1920–1949, in L. Chambers and E.-A. Montigny (eds) *Family Matters: Papers in post-confederation Canadian family history*, pp. 475–97. Toronto: Canadian Scholars Press.

Macdonald, B. and Rich, C. (1984) *Look Me in the Eye*, London: The Women's Press.

McGwin, G., Chapman, V., Curtis, J. and Rousculp, T. (2000) Fire fatalities in older people, *Journal of the American Geriatric Society*, 47: 1307–11.

McHugh, K. (2003) Three faces of ageism: society, image and place, *Ageing and Society*, 23: 165–85.

McKee, K., Harrison, G. and Lee, K. (1999) Activity, friendship and wellbeing in residential settings for older people, *Aging and Mental Health*, 3, 2: 143–52.

McKeever, Patricia (1999) Between women: nurses and family caregivers, *Canadian Journal of Nursing Research*, 30, 4: 185–91.

McKeever, Patricia (2001a) The 'hitting home' project: the home as a site for long-term care. Co-investigators: Jan Angus, Alfred Dolan, Isabel Dyck, Joan Eakin, Kim England, Denise Costaldo, Toronto: Home Care Evaluation and Research Centre, University of Toronto.

McKeever, Patricia (2001b) *'Hitting Home': The home as a locus of long-term care*, New York: Association of American Geographers Annual Meetings.

McKeever, P. and Coyte, P. (1999) Place in health care: sites, roles, rights and responsibilities. Report prepared under the auspices of a SSHRC/CHSRF Health Institute Design Grant.

McKeganey, Neil (1989) The role of home help organisers, *Social Policy and Administration*, 23, 2: 171–88.

McKeown, T. (1976) *The Modern Rise of Population*, London: Edward Arnold.

Maclean, A. (1986) *Night Falls on Ardnamurchan: The twilight of a crofting family*, Harmondsworth: Penguin.

Maddox, G. L. (2001) Housing and living arrangements: a transactional perspective, in R. H. Binstock and L. K. George (eds) *Handbook of Aging and the Social Sciences* (5th ed.), pp. 426–43, San Diego, CA: Academic Press.

Marans, R. W., Hunt, M. E. and Vakalo, K. L. (1984) Retirement communities, in I. Altman, M. P. Lawton and J. F. Wolwhill (eds) *Elderly People and the Environment*, pp. 57–93, New York: Plenum.

Marcus, Cooper C. (1995) *House as a Mirror of Self: Exploring the deeper meaning of home.* Berkeley, CA: Conari Press.

Marottoli, R. A., Mendes de Leon, C. F. Glass, T. A., Williams, C. S., Cooney, L. M., Berkman, L. F. and Tinetti, M. E. (1997) Driving cessation and increased depressive symptoms: prospective evidence from the New Haven EPESE, *Journal of the American Geriatrics Society*, 45: 202–6.

Marottoli, R. A., Mendes de Leon, C. F., Glass, T. A., Williams, C. S. Cooney, L. M. and Berkman, L. F. (2000) Consequences of driving cessation: decreased out-of-home activity levels, *Journal of Gerontology: Social Sciences* 55B, 6, S334–S340.

Marshall, M. (2001) Care settings and the care environment, in C. Cantley (ed.) *A Handbook of Dementia Care*, pp. 173–85, Buckingham: Open University Press.

Martin, L. G. (1989) Living arrangements of the elderly in Fiji, Korea, Malaysia, and the Philippines, *Demography*, 26, 4: 627–43.

Massey, D. (1994) *Space, Place and Gender*, Cambridge: Polity Press.

Masud, T. and Morris, R. O. (2001) Epidemiology of falls, *Age and Ageing*, 30, S4: 3–7.

Matthews, Anne Martin and Campbell, Lori (1995) Gender roles: employment and informal care, in S. Arber and J. Ginn (eds) *Connecting Gender and Ageing: A sociological approach*, pp. 139–43, Buckingham and Philadelphia: Open University Press.

Means, R. and Smith, R. (1998) *From Poor Law to Community Care*, Bristol: The Policy Press.

Mende, S. (2000) *Ageing in My Own Place*, London: HelpAge International.

Mermelstein, J. and Sundet, P. A. (1998) Rural social work is an anachronism: the perspective of twenty years of experience and debate, in L. Ginsberg (ed.) *Social Work in Rural Communities* (3rd edn), pp. 63–80, Virginia: Council on Social Work Education.

Merrett, C. D. (2001) Declining social capital and nonprofit organizations: consequences for small towns after welfare reform, *Urban Geography*, 22: 407–23.

Messecar, Deborah C. (2000) Factors affecting caregivers' ability to make environmental modifications, *Journal of Gerontological Nursing*, 26, 12: 32–42.

Miller, Baila and Cafasso, Lynda (1992) Gender differences in caregiving: fact or artefact? *The Gerontologist* 32: 498–507.

Milligan, Christine (2000) 'Bearing the burden': towards a restructured geography of caring, *Area*, 32, 1: 49–58.

Milligan, Christine (2001) *Geographies of Care: Space, place and the voluntary sector*. Aldershot and Burlington: Ashgate.

Mills, C. W. (1940) Situated actions and vocabularies of motive, *American Sociological Review*, 5: 904–13.

Minkler, M., Schauffler, H. and Clements-Nolle, K. (2000) Health promotion for older Americans in the 21st century, *American Journal of Health Promotion*, 14, 6: 371–9.

Miskelly, F. G. (2001) Assistive technology in elderly care, *Age and Ageing*, 30: 455–8.

Mitsch Bush, D. and Simmons, R. (1981) Socialization process over the life course, in M. Rosenberg and R. Turner (eds) *Social Psychology: Social perspectives*, pp. 133–64, New York: Basic Books.

Montgomery, Rhonda and McGlinn Datwyler, Mary (1990) Women and men in the caregiving role, *Generations*, 14, 3: 34–68.

Moody, H. (1993) Overview: what is critical gerontology and why is it important? in T. Cole, W. Achenbaum, P. Jakobi and R. Kastenbaum (eds) *Voices and Visions of Aging: Toward critical gerontology*, pp. xv–xxxvii, New York: Springer Publishing Company.

Moore, Eric, Rosenberg, Mark W. and McGuinness, Donald (1997) *Growing Old in Canada: Demographic and geographic perspectives*, Toronto: ITP Nelson.

Moore, Michael J., Zhu, Carolyn and Clipp, Elizabeth (2001) Informal costs of dementia care: estimates from the national longitudinal caregiver study, *Journal of Gerontology: Social Sciences* 56B, 4: S219–S228.

Moos, R. H. and Lemke, S. (1996) *Evaluating Residential Facilities*, London: Sage.

Mormont, M. (1990) What is rural? Or how to be rural: towards a sociology of the rural, in T. Marsden, R. Lowe and S. Whatmore (eds) *Rural Restructuring*, pp. 21–44, London: David Fulton Press.

Morris, Marika, Robinson, Jane, Simpson, Janet, Galey, Sherry, Kirby, Sandra, Martin, Lise and Muzychka, Martha (1999) *The Changing Nature of Home Care and Its Impact on Women's Vulnerability to Poverty*, Kingston: Canadian Research Institute for the Advancement of Women (CRIAW).

Morris, J. (2001) *Trieste and the Meaning of Nowhere*, London: Faber and Faber.

Moss, P. (1997) Negotiating spaces in home environments: older women living with arthritis, *Social Science and Medicine*, 45: 23–33.

Mowl, G., Pain, R. and Talbot, C. (2000) The ageing body and the homespace, *Area*, 32, 2: 189–97.

Moxon, S., Lyne, K., Sinclair, I., Young, P. and Kirk, C. (2001) Mental health in residential homes: a role for care staff, *Ageing and Society*, 21: 71–93.

Murdoch, J. and Pratt, A. (1993) Rural studies: modernism, postmodernism and the 'post-rural,' *Journal of Rural Studies*, 9: 411–28.

Mutchler, Jan E. and Bullers, Susan (1994) Gender differences in formal care use in later life, *Research on Aging*, 16, 3: 235–50.

Myer, P. (1982) *Aging in Place: Strategies to help elderly stay in revitalizing neighborhoods*, Washington, DC: The Conservation Foundation.

Myerhoff, B. (1978) *Number Our Days*, New York: Simon and Schuster.

Myerhoff, B. (1980) *Number Our Days*, Touchstone, NY: Simon and Schuster.

Myerhoff, B. (1992) *Remembered Lives: The work of ritual, storytelling and growing older*, Ann Arbor: University of Michigan Press.

Myerhoff, B. (2000) A death in due time: conviction, order, and continuity in ritual drama, in J. Gubrium and J. Hostein (eds) *Aging and Everyday Life*, pp. 417–39, Oxford: Blackwell Publishers.

Myers, A. M. and Huddy, L. (1985) Evaluating physical capabilities in the elderly: the relationship between ADL self-assessments and basic abilities, *Canadian Journal on Aging*, 4: 189–200.

Nahemow, L. (2000) The ecological theory of aging: Powell Lawton's legacy, in R. L. Rubinstein, M. Moss and M. H. Kleban (eds) *The Many Dimensions of Aging*, pp. 22–40, New York: Springer.

Naylor, David (1997) Building a sustainable health care system, *Canadian Speeches*, 11, 3: 25–31.

Netten, A., Darton, R., Bebbington, A. and Brown, P. (2001) Residential or nursing home care? The appropriateness of placement decisions, *Ageing and Society*, 21: 2–23.

Netten, A. (1993) *A Positive Environment? Physical and social influences on people with senile dementia in residential care*, Aldershot: Ashgate.

Neugarten, B. L. (1975) *Middle Age and Aging*, Chicago: University of Chicago Press.

New Zealand Ministry of Social Policy (2001) *The New Zealand Positive Ageing Strategy*. Retrieved 14 November 2002 from http://www.msp.govt.nz/keyinitiatives/positiveageing.html.

Newton, N. A., Brauer, D., Gutmann, D. L. and Grunes, J. (1986) Psychodynamic therapy with the aged, in T. L. Brink (ed.) *Clinical Gerontology*, New York: Hawthorn Press.

Noelker, Linda and Bass, David (1989) Home care for elderly persons: linkages between formal and informal caregivers, *Journal of Gerontology, Social Sciences*, 44, 2: S63–S69.

Noro, A. and Aro, S. (1997) Returning home from residential care? Patients' p and their determinants, *Ageing and Society*, 17: 305–21.

O'Rourke, N., MacLennan, R., Hadjistavropoulos, T. and Tuokko, H. (2000) Longitudinal examination of the functional status of older adults: successful aging within a representative Canadian sample, *Canadian Journal on Aging*, 19, 4: 441–55.

Office of Population Censuses and Surveys (1985) *National Population Projections*, London: HMSO.

Ogawa, N., Retherford R. D. and Saito, Y. (2003) Caring for the elderly and holding down a job: how are women coping in Japan? *Asia-Pacific Population and Policy* 65: 1–3.

Okumoto, Y., Koyama, E., Matsubara, H., Nakano, T. and Nakamura, R. (1998) Sleep improvement by light in a demented aged individual, *Psychiatry and Clinical Neuro-science*, 52, 2: 194–6.

Okun, M. A. and Keith, V. M. (1998) Effects of positive and negative social exchanges with various sources on depressive symptoms in younger and older adults, *Journal of Gerontology* 53B, 4–20.

Older Women's Network and Registered Nurses Association of Ontario (1998) Joint statement: the impact of health care restructuring on older women, RNAO.

Oldman, C. and Quilgars, D. (1999) The last resort? Revisiting ideas about older people's living arrangements, *Ageing and Society*, 19: 363–84.

Oliver, D. (2002) Bed falls and bed rails. What should we do? *Age and Ageing*, 31: 415–18.

Oliver, D., Hopper, A. and Seed, P. (2000) Do hospital fall prevention programs work? A systematic review, *Journal of the American Geriartric Society*, 48: 1679–89.

Olshanky, S. J. and Ault, A. B. (1986) The fourth stage of the epidemiologic transition: the age of delayed degenerative diseases, *The Milbank Quarterly*, 64, 3: 355–91.

Omran, A. R. (1971) The epidemiologic transition: a theory of the epidemiology of population change, *The Milbank Memorial Fund Quarterly*, 49, 4: 509–38. Reprinted in the *Bulletin of the World Health Organization* (2001) 79, 2 with a commentary by J. C. Caldwell.

Organisation for Economic Co-operation and Development (OECD) (2000) *Reforms for an Ageing Society*, Paris: OECD.

Oswald, F. (1996) *Hier bin ich zu Hause: Zur Bedeutung des Wohnens: Eine empirische Studie mit gesunden und gehbeeintrachtigten Älteren* [On the meaning of home: an empirical study with healthy and mobility-impaired elders], Regensburger: Roderer.

Pain, R., Mowl, G. and Talbot, C. (2000) Difference and the negotiation of 'old age', *Environment and Planning D: Society and space*, 18: 377–93.

Palmore, E. (1999) *Ageism: Negative and Positive*, New York: Springer.

Panelli, R., Stolte, O. and Bedford, R. D. (2003) The reinvention of Tirau: landscape as a record of change and culture, *Sociologia Ruralis* (under review).

Parker, K. and Miles, S. (1997) Death caused by bed rails, *Journal of the American Geriatric Society*, 45: 797–802.

Parmelee, P. A. and Lawton, M. P. (1990) Design of special environments for the elderly, in J. E. Birren and K. W. Schaie (eds) *Handbook of Psychology and Aging*, 3rd edn), pp. 464–87, New York: Academic Press.

Parr, H. (2002) Medical geography: diagnosing the body in medical and health geography, 1999–2000, *Progress in Human Geography*, 26, 2: 240–51.

Parr, H. (2003) Medical geography: care and caring, *Progress in Human Geography*, 27, 2: 212–21.

Passini, R., Pigot, H., Rainville, C. and Tetreault, M. H. (2000) Wayfinding in a nursing home for advanced dementia of the Alzheimer's type, *Environment and Behavior*, 32, 5: 684–710.

Passini, R., Rainville, C., Marchand, N. and Joanette, Y. (1998) Wayfinding and dementia: some research findings and a new look at design, *Journal of Architectural Planning and Research*, 15, 2: 133–51.

Peace, S. M. (1977) The elderly in an urban environment: a study of spatial mobility in Swansea, unpublished Ph.D. thesis, University College of Swansea, University of Wales.

Peace, S. M. (2002) The role of older people in social research, in A. Jamieson and

C. Victor (eds) *Researching Ageing and Later Life*, pp. 226–44, Buckingham: Open University Press.

Peace, S. and Holland, C. (eds) (2001a) *Inclusive Housing in an Ageing Society: Innovative approaches*, Bristol: The Policy Press.

Peace, S. and Holland, C. (2001b) Homely residential care: a contradiction in terms?, *Journal of Social Policy*, 30, 3: 393–410.

Peace, S., Hall, J. F. and Hamblin, J. R. (1979) *The Quality of Life of the Elderly in Residential Care*, London: Survey Research Unit, Polytechnic of North London.

Peace, S., Holland, C. and Kellaher, L. (2003) *Environment and Identity in Later Life: A cross-setting study*, end of award report to Economic and Social Research Council, Swindon.

Peace, S., Holland, C. and Kellaher, L. (eds) (forthcoming) *Environment and Identity in Later Life*, Maidenhead: Open University Press.

Peace, S. M., Kellaher, L. and Willcocks, D. (1992) *A Balanced Life: A consumer study of residential life in one hundred local authority old people's homes*, Research Report no. 14, Survey Research Unit, London: Polytechnic of North London.

Peace, S., Kellaher, L. and Willcocks, D. (1997) *Re-evaluating Residential Care*, Buckingham: Open University Press.

Pearlin, L. I. and Skaff, M. M. (1995) Stressors in adaptation in later life, in M. Gatz (ed.) *Emerging Issues in Mental Health and Aging*, Washington: American Psychological Association.

Percival, J. (2002) Domestic spaces: uses and meanings in the daily lives of older people, *Ageing and Society*, 22: 729–49.

Perkins, Harvey C. and Thorns, David C. (1999) House and home and their interaction with changes in New Zealand's urban system, households and family structures, *Housing Theory and Society*, 16, 3: 124–35.

Persson, L. O. (1992) Rural labour markets meeting urbanisation policy and planning, in T. Marsden, P. Lowe and S. Whatmore (eds) *Labour and Locality: Uneven development and the rural labour process*, London: David Fulton Press.

Phillips, D. R. (1993) Urbanization and human health, *Parasitology*, 106, S93–S107.

Phillips, D. R. (1999) The importance of the local environment in the lives of urban elderly people, in D. R. Phillips and A. G. O. Yeh (eds) *Environment and Ageing: Environmental policy, planning and design for elderly people in Hong Kong*, pp. 15–35, Hong Kong: Centre of Urban Planning and Environmental Management, University of Hong Kong.

Phillips, D. R. (ed.) (2000) *Ageing in the Asia-Pacific Region*, London: Routledge.

Phillips, D. R. and Chan, A. C. M. (eds) (2002) *Ageing and Long-term Care: National policies in the Asia–Pacific*, Singapore: ISEAS and Ottawa: IDRC.

Phillips, D. R. and Verhasselt, Y. (eds) (1994) *Health and Development*, London: Routledge.

Phillips, D. R. and Vincent, J. (1986a) Private residential accommodation for the elderly: geographical aspects of developments in Devon, *Transactions of the Institute of British Geographers*, NS 11: 155–73.

Phillips, D. R. and Vincent, J. (1986b) Petit Bourgeois Care: private residential care for the elderly, *Policy and Politics*, 14, 2: 189–208.

Phillips, D. R. and Vincent, J. (1988) Privatising residential care for the elderly: the geography of developments in Devon, England, *Social Science and Medicine*, 26, 1: 37–47.

Phillips, D. R. and Yeh, A. (eds) (1999) *Environment and Ageing: Environmental policy, planning and design for elderly people in Hong Kong*, Hong Kong: Centre of Urban Planning and Environmental Management, University of Hong Kong.

Phillips, D. R., Vincent, J. and Blacksell, S. (1987a) Spatial concentration of residential homes for the elderly: planning responses and dilemmas, *Transactions of the Institute of British Geographers*, NS 12: 73–83.

Phillips, D. R., Vincent, J. and Blacksell, S. (1987b) The development of private residential accommodation for the elderly: the case of Devon, England, *Espace Populations Sociétés*, 1: 235–48.

Phillips, D. R., Vincent, J. and Blacksell, S. (1988) *Home From Home? Private residential care for elderly people*, Sheffield: Joint Unit for Social Science Research.

Phillips, D. R., Siu, O. L., Yeh, A. G. O. and Cheng, K. H. C. (2004) Factors influencing older persons' residential satisfaction in big and densely populated cites in Asia: a case study in Hong Kong, *Ageing International*, 29, 1: 46–70.

Phillipson, C. (2003) Globalisation and the future of ageing: developing a critical gerontology, *Sociological Research Online*, 8, 4, http://www.socresonline.org.uk/8/4/phillipson.html.

Philo, C. (2000) Post-asylum geographies: an introduction, *Health and Place*, 6, 3: 135–6.

Pickard, Susan, Jacobs, Sally and Kirk, Susan (2003) Challenging professional roles: lay carers' involvement in health care in the community, *Social Policy and Administration*, 37, 1: 82–96.

Pinch, S. P. (1980) Local overview and case study in London, in D. Herbert and D. Smith (eds) *Social Problems and the City*, Oxford: Oxford University Press.

Pinfold, V. (2000) Building up safe havens all around the world: users' experiences of living in the community with mental health problems, *Health and Place*, 6, 3: 201–12.

Pinquart, M. and Sorensen, S. (2001) Influences on loneliness in older adults: a meta-analysis, *Basic and Applied Social Psychology*, 23, 4: 245–66.

Pinto, M. R., Medici, D. S., Zlotnicki, A., Bianchi, A., Sant, C. A. and Napoli, C. (1997) Reduced visual acuity in elderly people: the role of ergonomics and gerontechnology; *Age and Ageing*, 26: 339–44.

Poland, B., Green, L. and Rootman, I. (2000) *Settings for Health Promotion: Linking theory and practice*, Thousand Oaks, CA: Sage Publications.

Polder, J. J., Bonneux, L., Meerding, W. J. and Van Der Maas, P. J. (2002) Age-specific increases in health care costs, *European Journal of Public Health*, 12, 1: 57–62.

Powell, J. and Longino, C. (2001) Towards the postmodernization of aging: the body and social theory, *Journal of Aging and Identity*, 6: 199–207.

Prime Minister's Science, Engineering and Innovation Council (2003) *Promoting Healthy Ageing in Australia*, Canberra: PMSEIC.

Prince, M. J., Harwood, R. H., Blizard, R. A., Thomas, A. and Mann, H. (1997) Social support deficits, loneliness and life events as risk factors for depression in old age, the Gospel Oak Project VI, *Psychological Medicine*, 27: 323–32.

Proshansky, H. M., Fabian, A. K. and Kaminoff, R. (1983) Place-identity: physical world socialization of the self, *Journal of Environmental Psychology*, 46: 1097–1108.

Qureshi, Hazel (1990) A research note on the hierarchy of obligations among informal caregivers – a response to Finch and Mason, *Ageing and Society*, 10: 455–8.

Qureshi, Hazel and Walker, Alan (1989) *The Caring Relationship: Elderly people and their families*, Basingstoke: Macmillan.

Ramp, W. (1999) Rural health: context and community, in W. Ramp, J. Kulig, I. Townshend and V. McGowan (eds) *Health in Rural Settings: Contexts for Action*, pp. 1–13, Alberta: University of Lethbridge.

Ranzijn, R. (2002) Towards a positive psychology of ageing: potentials and barriers, *Australian Psychologist*, 37, 2: 79–85.

Ratzan, S. C., Filerman, G. L. and LeSar, J. W. (2000) Attaining global health: challenges and opportunities, *Population Bulletin*, 55, 1, Washington DC: Population Reference Bureau.
Ray, R. (2000) *Beyond Nostalgia: Aging and life-story writing*, Charlottesville: University Press of Virginia.
Raynes, N. V. (1998) Involving residents in quality specification, *Ageing and Society*, 18: 65–78.
Readt, R. E. and Ponjaert-Kristoffersen, P. (2000) Can strategic and tactical compensation reduce crash risk in older drivers? *Age and Ageing*, 29: 517–21.
Redfoot, R. (1987) 'On the separatin' place': social class and relocation among older women, *Social Forces*, 66: 486–500.
Reed, J. and Roskell-Payton, V. (1996) Constructing familiarity and managing the self: ways of adapting to life in nursing and residential homes for older people, *Ageing and Society*, 16: 543–60.
Reed, J. and Royskell-Payton, V. (1997) Understanding the dynamics of life in care homes for older people: implications for de-institutionalizing practice, *Health and Social Care in the Community*, 5, 4: 261–8.
Reed, J., Cook, G., Sullivan, A. and Burridge, C. (2003) Making a move: care-home residents' experiences of relocation, *Ageing and Society*, 23: 225–41.
Reed-Danahay, D. (2001) 'This is your home now!' Conceptualizing location and dislocation in a dementia unit, *Qualitative Research*, 1, 1: 47–63.
Regnier, V. A. (1974) Matching an older person's cognition with their use of neighborhood areas, in D. H. Carson (ed.) *Man–Environment Interactions* (vol. 3), pp. 19–40, New York: Halstead.
Reich, J. W. and Zautra, A. J. (1984) Daily event causation: an approach to elderly life quality, *Journal of Community Psychology*, 12: 312–22.
Reimer, B., Ricard, I. and Shaver, F. M. (1992) Rural deprivation: a preliminary analysis of census and tax family data, in R. Bollman (ed.) *Rural and Small Town Canada*, Ottawa: Thompson Educational Publishing,.
Relph, E. (1976) *Place and Placelessness*, London: Pion.
Rice, J. J. and Prince, M. J. (2001) *Changing Politics of Canadian Social Policy*, Toronto: University of Toronto Press.
Robson, D., Nicholson, A. M. and Baker, N. (1997) *Homes for the Third Age: A design guide for extra care sheltered housing*, London: E. and F. N. Spon.
Rodeheaver, D. and Datan, N. (1988) The challenge of double jeopardy: toward a mental health agenda for aging women, *American Psychologist*, 43: 648–54.
Rodin, J. (1986a) Health, control and aging, in M. M. Baltes and P. B. Baltes (eds) *The Psychology of Control and Aging*, Hillsdale, NJ: LEA.
Rodin, J. (1986b) Aging and health: effects of the sense of control, *Science*, 233: 1271–6.
Rodin, J. (1990) Control by any other name: definitions, concepts, and processes, in J. Rodin, C. Schooler and K. W. Schaie (eds) *Self-Directedness: Cause and effects throughout the life course*, pp. 1–17, Hillsdale, NJ: Erlbaum.
Rodin, J. and Langer, E. J. (1977) Long-term care effects on a control-relevant intervention with the institutionalized aged, *Journal of Personality and Social Psychology*, 12: 897–902.
Rogers, R. G. and Hackenberg, R. (1987) Extending epidemiologic transition theory: a new stage, *Social Biology* 34, 3–4: 234–43.
Rolls, Elizabeth (1992) Do the health visitor's professional training and bureaucratic responsibilities separate her from the women she is serving? *Women's Studies International Forum*, 15, 3: 397–404.

Rose, Gillian (1993) *Feminism and Geography: The limits of geographical knowledge*, Minneapolis: University of Minnesota Press.

Rose, Gillian (2001) *Visual Methodologies*, London: Sage.

Rose, Hilary and Bruce, Errolyn (1995) Mutual care but differential esteem: caring between older couples, in S. Arber and J. Ginn (eds) *Connecting Gender and Ageing: A sociological approach*, pp. 114–28, Buckingham and Philadelphia: Open University Press.

Rosenberg, M. (1965) *Society and the Adolescent Self-image*, Princeton, NJ: Princeton University Press.

Rosenberg, M. and Moore, E. (2001) Canada's elderly population: the challenges of diversity, *Canadian Geographer*, 45: 145–50.

Rose-Rego, Sharon K., Strauss, Milton E. and Smyth, Kathleen A. (1998) Differences in the perceived well-being of wives and husbands caring for persons with Alzheimer's disease, *The Gerontologist*, 38, 2: 224–30.

Rotter, J. B. (1954) *Social Learning and Clinical Psychology*, Englewood Cliffs, NJ: Prentice-Hall.

Rotter, J. B. (1990) Internal versus external control of reinforcement, *American Psychologist*, 45: 489–93.

Rowe, J. W. and Kahn, R. L. (1987) Human aging: usual and successful, *Science*, 237, 4811: 143–9.

Rowe, J. W. and Kahn, R. L. (1997) Successful aging, *The Gerontologist*, 37, 4: 433–40.

Rowe, J. W. and Kahn, R. L. (1998) *Successful Aging*, New York: Pantheon.

Rowe, J. W. and Kahn, R. L. (1999) The future of aging, *Contemporary Longterm Care*, 22, 2: 36–8.

Rowles, Graham D. (1978) *Prisoners of Space? Exploring the geographic experience of older people*, Boulder, CO: Westview.

Rowles, Graham D. (1983a) Place and personal identity in old age: observations from Appalachia, *Journal of Environmental Psychology* 3: 299–313.

Rowles, Graham D. (1983b) Geographical dimensions of social support in rural Appalachia, in G. D. Rowles and R. J. Ohta (eds) *Aging and Milieu: Environmental perspectives on growing old*, pp. 111–30, New York: Academic Press.

Rowles, Graham D. (1986) The geography of ageing and the aged: towards an integrated perspective, *Progress in Human Geography*, 10: 511–39.

Rowles, Graham D. (1991) Beyond performance: being in place as a component of occupational therapy, *American Journal of Occupational Therapy*, 45: 111–30.

Rowles, Graham D. (1993) Evolving images of place in aging and 'aging in place', *Generations*, 17, 2: 65–70.

Rowles, Graham D. (2000) Habituation and being in place, *Occupational Therapy Journal of Research*, 20 (Supplement 1): 52S–67S.

Rowles, Graham and Ohta, R. J. (eds) (1983) *Aging and Milieu: Environmental perspectives on growing old*, New York: Academic Press.

Rowles, Graham D. and Watkins, J. F. (2003) History, habit, heart, and hearth: on making spaces into places, in K. W. Schaie, H. W. Wahl, H. Mollenkopf and F. Oswald (eds) *Aging Independently: Living Arrangements and Mobility*, New York: Springer.

Rubinstein, R. L. (1986) *Singular Paths: Old men living alone*, New York: Columbia.

Rubinstein, R. L. (1987) The significance of personal objects to older people, *Journal of Aging Studies*, 4: 131–48.

Rubinstein, R. L. (1989) The home environments of older people: a description of the psycho-social processes linking person to place, *Journal of Gerontology: Social Sciences*, 44, 45–56.

Rubinstein, R. L. (1990a) Culture and disorder in the home care experience, in J. Gubrium and A. Sankar (eds) *The Home Care Experience*, pp. 37–58, Newbury Park, CA: Sage.

Rubinstein, R. L. (1990b) Personal identity and environmental meaning in later life, *Journal of Aging Studies*, 4: 131–48.

Rubinstein, R. L. (1999) The importance of including the home environment in assessment of frail older persons, *Journal of the American Geriatric Society*, 47: 111–12.

Rubinstein, R. L. and De Medeiros, K. (2004) Ecology and the aging self, in H. Wahl, R. J. Scheidt and P. G. Windley (eds) Aging in context: socio-physical environments, *Annual Review of Gerontology and Geriatrics* (vol. 23, 2003, pp. 59–84), New York: Springer Publishing Company.

Rubinstein, R. L. and Parmelee, P. A. (1992) Attachment to place and the representation of the life course by the elderly, in I. Altman and S. M. Low (eds) *Place Attachment*, pp. 139–60, New York and London: Plenum Press.

Rubinstein, R. L. and Powers, C. (1999) *Falls and Mobility Problems: Potential quality indicators and literature review (the ACOVE project)*, Santa Monica, CA: Rand Corporation.

Ruddick, William (1994) Transforming homes and hospitals, *Hastings Center Report*, 24, 5: S11–S14.

Ryke, E., Niba, T. and Strydom, H. (2003) Perspectives of elderly blacks on institutional care, *Social Work*, 39, 2: 139–48.

Sceats, S. (2003) Glimpses of the cadaver: fiction and the aging body. Paper presented at the The Flesh Made Text: Bodies, Theories, Cultures in the Post-Millennial Era, Aristotle University of Thessaloniki, Greece.

Schaie, K. W. (1996) *Intellectual Development in Adulthood: The Seattle longitudinal study*, New York: Cambridge University Press.

Schaie, K. W. and Willis, S. L. (1991) Adult personality and psychomotor performance: cross-sectional and longitudinal analyses, *Journal of Gerontology: Psychological series* 46: P275–P284.

Scheidt, R. J. and Norris-Baker, C. (2004) The general ecological model revisited: evolution, current status, and continuing challenges, in H. Wahl, R. J. Scheidt and P. G. Windley (eds) Aging in context: socio-physical environments, *Annual Review of Gerontology and Geriatrics* (vol. 23, 2003, pp. 34–58), New York: Springer Publishing Company.

Schulz, R. (1976) Effects of control and predictability on the physical and psychological well-being of the institutionalized aged, *Journal of Personality and Social Psychology*, 33: 563–73.

Schulz, R. and Hanusa, B. H. (1978) Long-term effects of control and predictability-enhancing interventions: findings and ethical issues, *Journal of Personality and Social Psychology*, 36: 1194–201.

Schulz, R. and Heckhausen, J. (1999) Aging, culture and control: setting a new research agenda, *Journals of Gerontology* 54B, 3: 139–45.

Scott, Anne and Wenger, G. Clare (1995) Gender and social support networks in later life, in S. Arber and J. Ginn (eds) *Connecting Gender and Ageing: A sociological perspective*, pp. 158–72, Buckingham and Philadelphia: Open University Press.

Scully, Tom, van der Walde, Lambert and Lindstrom, Laurel (2003) CMS Health Care Industry Market Update – Home Health, Centers for Medicare and Medicaid Services (CMS) and the United States Department of Health and Human Services (HHS): 25.

Seabrook, J. and Roberts, Y. (1981) Mrs Thatcher's heartland, *Observer Magazine*, 8 February: 26–33.

Secretary of State for Health (1999) *Saving Lives: Our healthier nation*, London: The Stationery Office.

SeniorWatch (2003) http://www.seniorwatch.de/swa/. Website accessed 15 October 2003.

Sheehy, G. (1976) *Passages: Predictable crises of adult life*, New York: Dutton.

Sherwood, S., Greer, D. S., Morris, J. N. and, Mor, V. (1981) *Alternative to Institutionalization: The Highland Heights experiment*, Cambridge, MA: Ballinger.

Sillitoe, A. (2002) *Birthday*, London: Flamingo.

Silverman, P. (1987) *The Elderly as Modern Pioneers*, Bloomington: Indiana University Press.

Simpson, M. (2002) 101 and Elizabeth switches on at last to the 21st century, *Aberdeen Evening Express*, 13 December: 10.

Siu, O. L., Phillips, D. R., Yeh, A. and Cheng, K. H. C. (2003) The mediating and moderating effects of self-esteem and feelings of control on the influence of housing environment on psychological well-being of older persons in Hong Kong, Hong Kong: Lingnan University,.

Sixsmith, A. (1990) The meaning and experience of home in later life, in B. Bytheway and J. Johnson (eds) *Welfare and the Ageing Experience*, pp. ??????, Aldershot: Gower.

Sixsmith, J. (1986) The meaning of home: an exploratory study of environmental experience, *Journal of Environmental Experience*, 6: 281–98.

Slaughter, V. (1997) *The Elderly in Poor Urban Neighbourhoods*, New York: Garland Publishing.

Slivinske, L. R. and Fitch, V. L. (1987) The effect of control enhancing intervention on the well-being of elderly individuals living in retirement communities, *The Gerontologist*, 27: 176–81.

Smith, G. C. and Ford, R. G. (1998) Geographical change in residential care provision for the elderly in England, 1988–1993, *Health and Place*, 4: 15–31.

Smith, J. (1994) *Transport and Welfare*, Salford: University of Salford, Transport Geography Study Group, Institute of British Geographers.

Smith, J. (2001) Life experience and longevity: findings from the Berlin Aging Study, *Zeitschrift für Erziehungswissenschaft*, 4: 577–99.

Smith, J. and Baltes, P. B. (1996) Aging from a psychological perspective: trends and profiles in old age, in K. U. Mayer and P. B. Baltes (eds) *Die Berliner Altersstudie* (vol. 1), pp. 221–50, Berlin, Germany, Akademieverlag.

Smith, R. E. (1993) *Psychology*, St Paul, MN: West Publishing Company.

Soja, E. (1989) *Postmodern Geographies*, London: Verso.

Sontag, S. (1977) *On Photography*, Toronto: McGraw-Hill.

Stainer, L. (1999) Living alone in later life: a case study of geographical gerontology, *East Midland Geographer*, 21: 12–23.

Startzell, J. K., Owens, D. A., Mulfinger, L. M. and Cavanagh, P. R. (2000) Stair negotiation in older people: a review, *Journal of the American Geriatric Society*, 48: 567–8.

Steichen, E. (1955) *The Family of Man*, New York: Museum of Modern Art.

Steiner, A. (2001) Intermediate care: more than a nursing thing, *Age and Aging*, 30: 433–5.

Stevens, M., D'Arcy, C. and Bennett, N. (2001) Preventing falls in old people: impact on an intervention to reduce environmental hazards in the home, *Journal of the American Geriatric Society*, 49: 1442–7.

Stewart, A. L. and King, A. C. (1994) Conceptualizing and measuring quality of life in older populations, in R. P. Abeles, H. C. Gift and M. G. Ory (eds) *Aging and Quality of Life*, pp. 27–54, New York: Springer.

Stoller, Eleanor Palo (1988) Long-term care planning by informal helpers: likelihood of shared households and institutional placement, *Journal of Applied Gerontology*, 7, 1: 5–20.

Stoller, Eleanor Palo (1994) Teaching about gender: the experience of family care of frail elderly relatives, *Educational Gerontology*, 20: 679–97.

Stone, Robyn, Cafferata, Gail Lee and Sangl, Judith (1987) Caregivers of the frail elderly: a national profile, *The Gerontologist*, 27, 5: 616–26.

Stott, W. (1973) *Documentary Expression and Thirties America*, Oxford: Oxford University Press.

Strategic Injury Prevention Partnership (2001) *National Injury Prevention Plan: Priorities for 2001–2003*, Canberra, ACT: Department of Health and Aged Care, Commonwealth of Australia.

Suh, G.-H. and Shah, A. (2001) A review of the epidemiological transition in dementia – cross-national comparisons of the indices related to Alzheimer's disease and vascular dementia, *Acta Psychiatrica Scandinavica*, 104: 4–11.

Sundström, Gerdt and Tortosa, Maria Angeles (1999) The effects of rationing home-help services in Spain and Sweden: a comparative analysis, *Ageing and Society*, 19, 3: 343–61.

Swain, J., Finkelstein, V., French, S. and Oliver, M. (1993) *Disabling Barriers – Enabling Environments*, London: Sage.

Sykes, J. B. (1976) *The Concise Oxford Dictionary* (6th edn), Oxford: Clarendon Press.

Teather, E. and Argent, N. (1998) The social sustainability of rural and remote New South Wales, Australia: can Australia sustain its residents through the 'seven ages' of the lifecourse? in R. Epps, (ed.) *Sustainable Rural Systems in the Context of Global Change*, pp. 381–98, Armidale: School of Geography, University of New England.

Teeland, L. (1998) Home, sick: implications of health care delivery in the home, *Scandinavian Housing and Planning Research*, 15, 4: 271–82.

Templer, J. (1992) *The Staircase: Studies of hazards, falls and safer designs*, Cambridge, MA: MIT Press.

Thoits, P. A. (1995) Stress, coping and social support processes: where are we? What next? *Journal of Health and Social Behavior* 36 (Suppl.): 53–79.

Thompson, P., Itzin, C. and Abendstern, M. (1990) *I Don't Feel Old*, Oxford: Oxford University Press.

Tibbenham, A. (1985) *Private and Local Authority Care of the Elderly in Devon*, Exeter: Devon County Council.

Tideiksaar, R. (1989) Geriatric falls: assessing the cause, preventing recurrence, *Geriatrics*, 44: 57–64.

Tinetti, M. E. and Speechley, M. (1989) Prevention of falls among the elderly, *New England Journal of Medicine*, 320: 1055–9.

Tinetti, M. E., Richman, D. and Powell, L. (1990) Falls efficacy as a measure of fear of falling, *The Journals of Gerontology. Series B, Psychological sciences and social sciences*, 45: 239–43.

Tinetti, M. E., Doucette, J., Claus, E. and Marottoli, R. (1995) Risk factors for serious injury during falls by older persons in the community, *Journal of the American Geriatrics Society*, 43: 1214–21.

Tornstam, L. (1989) Gero-transcendence: a meta-theoretical reformulation of the disengagement theory, *Aging: Clinical and experimental research*, 1, 1: 55–63.

Townsend, P. (1962) *The Last Refuge: A survey of residential institutions and homes for the aged in England and Wales*, London: Routledge.

Troughton, M. (1999) Redefining 'rural' for the twenty-first century, in W. Ramp, J. Kulig, I. Townshend and V. McGowan (eds) *Health in Rural Settings: Contexts for action*, pp. 31–8, Alberta: University of Lethbridge.

Turner, B. (1995) Aging and identity: some reflections on the somatization of the self, in M. Featherstone and A. Wernick (eds) *Images of Aging: Cultural representations of later life*, pp. 245–60, London: Routledge.

Turrell, A. (2001) Nursing homes: a suitable alternative to hospital care for older people in the UK? *Age and Ageing*, 30, S3: 24–32.

Twigg, J. (1997) Deconstructing the 'social bath': help with bathing at home for old and disabled people, *Journal of Social Policy*, 26, 2: 211–32.

Twigg, J. (1999) The spatial ordering of care: public and private in bathing support at home, *Sociology of Health and Illness*, 21: 381–400.

Twigg, Julia (1999) The spatial ordering of care: public and private in bathing support at home, *Sociology of Health and Illness*, 21, 4: 381–400.

Twigg, Julia (2000a) *Bathing – The Body and Community Care*, London: Routledge.

Twigg, Julia (2000b) Carework as a form of bodywork, *Ageing and Society*, 20: 389–411.

Twigg, Julia and Karl Atkin (1994) *Carers Perceived: Policy and practice in informal care*, Buckingham and Philadelphia: Open University Press.

Twigger-Ross, C. L. and Uzzell, D. L. (1996) Place and identity processes, *Journal of Environmental Psychology*, 16: 205–20.

Ungerson, Clare (1997) Social politics and the commodification of care, *Social Politics*, 4, 3: 362–81.

United Nations (UN) (2002a) *Madrid International Plan of Action on Ageing*, Second World Assembly on Ageing, Madrid, United Nations; website: www.un.org/ageing/coverage/index.html.

United Nations (UN) (2002b) *World Population Ageing 1950–2050*, New York: United Nations.

United Nations (UN) (2004) Replacement migration: is it a solution to declining and ageing populations? Retrieved 24 July 2004 from http://www.un.org/esa/population/publications/migration/migration.htm.

United Nations, Population Division (UNPD) (2003), *World Population Prospects: The 2002 revision population database* (http://esa.un.org/unpp).

US Department of Health and Human Services (2000) *Healthy People 2010: Understanding and improving health.* Retrieved 21 January 2003 from http://www.health.gov/healthypeople/.

Valencia Forum (2002) Statement by Gary Andrews. Retrieved 1 December 2003 from http://www.un.org/ageing/coverage/vforumE.htm.

Van Beurden, E., Kempton, A., Sladden, T. and Garner, E. (1998) Designing an evaluation for a multiple-strategy community intervention: the North Coast Stay on Your Feet program, *Australian and New Zealand Journal of Public Health*, 22, 1: 115–19.

Van Gennep, A. (1960) *The Rites of Passage*, trans. Monika B. Vizedom and Gabrielle L. Caffee, Chicago: University of Chicago Press.

Vassallo, M., Amersey, R. A., Sharma, J. C. and Allen, S. C. (2000) Falls on integrated medical wards, *Gerontology*, 46: 158–62.

Veitch, R. and Arkkelin, D. (1995) *Environmental Psychology: An interdisciplinary perspective*, Englewood Cliffs, NJ: Prentice-Hall, Inc.

Vellas, B. J., Baumgartner, R. N., Romero, L. J., Wayne, S. J. and Garry, P. J. (1993) Incidence and consequences of falls in free-living healthy elderly persons, in

J. L. Albarede and P. J. Garry (eds) *Facts and Research in Gerontology* (vol. 7), pp. 131–41, New York: Springer.
Vellas, B., Cayla, F., Bocquet, H., dePemille, F. and Albarede, J. L. (1989) Prospective study of restriction of activity in old people after falls, *Age and Ageing*, 18: 47–51.
Victor, C. (1996) Old age in the inner city: an analysis of data from over 75 GP screening assessments, *Health and Place*, 2: 221–7.
Vincent, J., Tibbenham, A. and Phillips, D. R. (1988) Choice in residential care: myths and realities, *Journal of Social Policy*, 16: 435–55.
Vittoria, A. K. (1998) Preserving selves: identity work and dementia, *Research on Aging*, 20, 1: 91–136.
von Faber, M., Bootsma-van der Wiel, A., van Exel, E., Gussekloo, J., Lagaay, A., van Dongen, E., Knook, D., van der Geest, S. and Westendorp, R. (2001) Successful aging in the oldest old: who can be characterized as successfully aged? *Archives of Internal Medicine*, 161, 22: 2694–700.
VROM (Ministerie van Volkshuisvesting, Ruimtelijke Ordening en Milieubeheer) (1997) *Huisvesting van ouderen op her breukvlak van twee eeuwen*, Zoetermeer: VROM.
Wahl, H. W. (2001) Environmental influences on aging and behavior, in J. E. Birren and K. W. Schaie (eds) *Handbook of the Psychology of Aging*, 5th edn), pp. 215–37, San Diego, CA: Academic Press.
Wahl, H. W., Schilling, O., Oswald, F. and Heyl, V. (1999) Psychosocial consequences of age-related visual impairment: comparison with mobility-impaired older adults and long-term outcome, *Journals of Gerontology* 54B, 5: 304–16.
Walker, A. (1993) Community care policy: from consensus to conflict, in J. Bornat, P. Charmaine, D. Pilgrim and F. Williams (eds) *Community Care: A reader*, London: Macmillan.
Walsh, D. A., Krauss, I. K. and Regnier, V. A. (1981) Spatial ability, environmental knowledge and environmental use: the elderly, in L. S. Liben, A. H. Patterson and N. Newcombe (eds) *Spatial Representation and Behavior across the Life-Span: Theory and application*, pp. 321–57, New York: Academic Press.
Ward-Griffin, Catherine and Marshall, Victor (2003) Reconceptualising the relationship between 'public' and 'private' eldercare, *Journal of Aging Studies*, 17: 189–208.
Warnes, A. (ed.) (1982) *Geographical Perspectives on the Elderly*, Chichester: Wiley.
Warnes, A. (ed.) (1989) *Human Ageing and Later Life: Multidisciplinary perspectives*, London: Edward Arnold.
Warnes, A. (1990) Geographical questions in gerontology: needed directions for research, *Progress in Human Geography*, 14: 24–56.
Warnes, A. (1994) Cities and elderly people: recent population and distributional trends, *Urban Studies*, 31: 799–816.
Warnes, A. (1999) UK and western European late-age mortality: trends in cause-specific death rates, 1960–1990, *Health and Place*, 5: 111–18.
Warnes, A. and Ford, R. (1992) *The Changing Distribution of Elderly People: Great Britain, 1981–91*, Department of Geography and Age Concern Institute of Gerontology, Occasional Paper no. 37.
Warnes, A. and Law, C. M. (1984) The elderly population of Great Britain: locational trends and policy implications, *Transactions of the Institute of British Geographers*, NS 9: 37–59.
Weisman, G. D. and Moore, K. D. (2003) Visions and values: M. Powell Lawton and the philosophical foundations of environment-aging studies, in R. J. Scheidt and

P. G. Windley (eds) *Physical Environment and Aging: Critical contributions of M. Powell Lawton to theory and practice*, pp. 23–37, New York: The Harworth Press.
Weisz, J. R. (1983) Can I control it? The pursuit of veridical answers across the lifespan, *Lifespan Development and Behavior*, 5: 233–300.
White, R. W. (1959) Motivation reconsidered: the concept of competence, *Psychological Review*, 66: 297–323.
Wiles, Janine L. (2003a) Daily geographies of caregivers: mobility, routine, scale, *Social Science and Medicine*, 57, 7: 1307–25.
Wiles, Janine L. (2003b) Informal caregivers' experiences of formal support in a changing context, *Health and Social Care in the Community*, 11, 3: 189–298.
Wiles, Janine L. and Rosenberg, Mark W. (*in submission*) Informal privatisation and private informalisation: interactions among levels of government and geographic scale, *International Journal of Canadian Studies*.
Willcocks, D., Peace, S. and Kellaher, L. (1982) *The Residential Life of Old People: A study in 100 local authority homes*, Survey Research Unit, Polytechnic of North London.
Willcocks, D., Peace, S. and Kellaher, L. (1987) *Private Lives in Public Places: A research-based critique of residential life in local authority old peoples' homes*, London: Tavistock.
Williams, A. (1999) *Therapeutic Landscapes: The dynamic between place and wellness*, Lanham, MD: University Press of America.
Williams, Allison M. (1996) The development of Ontario's home care program: a critical geographical analysis, *Social Science and Medicine*, 42, 6: 937–48.
Williams, Allison M. (2001) Home care restructuring at work: the impact of policy transformation on women's labour, in I. Dyck, N. Davis Lewis and S. McClafferty (eds) *Geographies of Women's Health*, pp. 107–26, London: Routledge.
Williams, Allison M. (2002) Changing geographies of care: employing the concept of therapeutic landscapes as a framework in examining home space, *Social Science and Medicine*, 55, 1: 141–54.
Williams, Allison, Wagner, Susan and Buettner, Monic (2002) Labour process change: women's paid home care work in Saskatoon, *Centres of Excellence for Women's Health Research Bulletin*, 3, 1: 13–15.
Williams, R. (1990) *A Protestant Legacy: Attitudes to illness and death among older Aberdonians*, Oxford: Clarendon Press.
Wilson, O. J. (1995) Rural restructuring and agriculture–rural economy linkages: a New Zealand case study, *Journal of Rural Studies*, 11: 417–31.
Winston, T. (1998) 'The camera never lies': the partiality of photographic evidence, in *Image-based Research*, London: Falmer Press.
Wister, A. V. (1989) Environmental adaptation by persons in their later life, *Research on Aging*, 11: 267–91.
Wistow, G. (1995a) Aspirations and realities: community care at the crossroads, *Health and Social Care in the Community*, 3, 4: 227–40.
Wistow, G. (1995b) Coming apart at the seams, *Health Services Journal*, 1: 24–5.
Woodward, K. (1991) *Aging and Its Discontents: Freud and other fictions*, Bloomington: Indiana University Press.
World Health Organization (WHO) (1946) *Constitution*, New York: World Health Organization.
World Health Organization (WHO) (1952) Constitution of the World Health Organization, in *World Health Organization Handbook of Basic Documents*, 5th edn, Geneva: Palais des Nations.

World Health Organization (WHO) (1986) *Ottawa Charter for Health Promotion.* Paper presented at the First International Conference on Health Promotion, 21 November 1986, Ottawa, Canada.

World Health Organization (WHO) (2002) *Active Ageing: A policy framework,* Geneva: WHO. Retrieved 21 January 2003 from http://www.who.int/hpr/ageing/ActiveAgeingPolicyFrame.pdf.

Worrell, L., Bartlett, H. and Boldy, D. (2002) Conceptualising age friendly. Paper presented at International Federation on Ageing 6th Global Conference: Maturity Matters, Perth, WA, 27–30 October 2002.

Yamada, A. (2002) The evolving retirement income package: trends in adequacy and equality in nine OECD countries. *Labour Market and Social Policy Occasional Papers No. 63,* Paris: OECD.

Yassuda, M. S., Wilson, J. J. and von Mering, O. (1997) Driving cessation: the perspective of senior drivers, *Educational Gerontology,* 23: 525–38.

Youmans, E. G. (1969) Some perspectives on disengagement theory, *The Gerontologist,* 9, 254.

Young, Rosalie F. and Kahana, Eva (1989) Specifying caregiver outcomes: gender and relationship aspects of caregiving strain, *The Gerontologist,* 29: 660–6.

Zapf, M. (2001) Notions of rurality, *Rural Social Work,* Special Australian/Canadian Issue, December: 12–27.

Zarit, Steven H., Gaugler, Joseph E. and Jarrott, Shannon E. (1999) Useful services for families: research findings and directions, *International Journal of Geriatric Psychiatry,* 14, 3: 165–78.

Zingmark, A., Norberg, K. and Sandman, P. O. (1995) The experience of being at home throughout the life course: investigation of persons 2 to 102, *International Journal of Aging and Human Development,* 41, 1: 47–62.

INDEX

For Product Safety Concerns and Information please contact our EU
representative GPSR@taylorandfrancis.com
Taylor & Francis Verlag GmbH, Kaufingerstraße 24, 80331 München, Germany

www.ingramcontent.com/pod-product-compliance
Ingram Content Group UK Ltd.
Pitfield, Milton Keynes, MK11 3LW, UK
UKHW021008180425
457613UK00019B/849

* 9 7 8 0 4 1 5 4 8 1 6 5 6 *